The
original edition
of this book con-
tained oversized maps.
PDF files of these maps
can be downloaded from
www.elibron.com/maps

ARTHUR JOHN HAYES

THE SOURCE OF THE BLUE NILE

A RECORD OF A JOURNEY THROUGH THE SOUDAN TO LAKE TSANA IN WESTERN ABYSSINIA, AND OF THE RETURN TO EGYPT BY THE VALLEY OF THE ATBARA

WITH

A NOTE ON THE RELIGION, CUSTOMS, ETC. OF ABYSSINIA

AN ETIMOLOGICAL APPENDIX

BY

E.B. POULTON

Elibron Classics
www.elibron.com

Elibron Classics series.

ISBN 1-4021-5788-6 (paperback)
ISBN 1-4021-0392-1 (hardcover)

This Elibron Classics Replica Edition is an unabridged facsimile of the edition published in 1905 by Smith, Elder and Co., London.

THE SOURCE OF THE BLUE NILE

THE SOURCE OF THE BLUE NILE SHOWING LAKE IN THE DISTANCE.

[*Frontispiece.*
See p. 136.

THE SOURCE OF THE BLUE NILE

A RECORD OF A JOURNEY THROUGH THE SOUDAN TO LAKE TSANA IN WESTERN ABYSSINIA, AND OF THE RETURN TO EGYPT BY THE VALLEY OF THE ATBARA

WITH

A NOTE ON THE RELIGION, CUSTOMS, ETC. OF ABYSSINIA

BY

ARTHUR J. HAYES, L.S.A.(LOND.)

MEDICAL OFFICER, QUARANTINE OFFICE, SUEZ

AND

AN ENTOMOLOGICAL APPENDIX

BY

E. B. POULTON, LL.D., F.R.S.

HOPE PROFESSOR OF ZOOLOGY IN THE UNIVERSITY OF OXFORD

WITH ILLUSTRATIONS AND MAPS

LONDON

SMITH, ELDER & CO., 15, WATERLOO PLACE

1905

PRINTED BY
WILLIAM CLOWES AND SONS, LIMITED,
LONDON AND BECCLES.

PREFACE

In Africa the centre of interest shifts quickly, from Khartoum to the Cape, from the Congo to Morocco. Before now it has lain in Abyssinia, for Englishmen especially. It may be found there again. If so, the theatre of action will probably be the little known region of Western Abyssinia, and that district of the Anglo-Egyptian Nile Province which adjoins it.

Geographically, Western Abyssinia dominates the south-east of the Soudan. The Soudan, as every one in England knows now, is not a continuation of the Desert of Sahara, but a land that once flowed with milk and honey, and may again. It contains vast tracts of soil perfectly adapted for the cultivation of cotton. A hostile force descending from Abyssinia has the enormous advantage of moving from difficult into easy country with an open line of retreat into almost inaccessible mountains. An expedition from the Soudan, on the other hand, would be confronted, after traversing miles of uninhabited hilly wastes, by the necessity of forcing its way up mule-paths winding among precipices.

There is no reason why peace should not be permanently established between Egypt and Ethiopia, if the Abyssinian slave-raids are stopped. But the changes and chances of international politics bring about strange consequences. Rumours, not without foundation, have been circulated recently of new engagements entered into by the Negus giving far-reaching concessions to Americans. Other Powers are busy, and a diplomatic —and spectacular—mission started lately from Berlin for Addis Abbiba. There is room in the country for all nations to find commercial opportunities. But if influences hostile to Great Britain became dominant in Western Abyssinia, a danger to the Soudan—and not to the Soudan only—would have arisen, the seriousness of which few people at home, perhaps, rightly realize. I make no further apology for bringing some account of a journey from Khartoum to Lake Tsana before the public.

My heartiest thanks are due to my friend, Mr. Godfrey Burchett, without whose aid in preparing for publication the rough notes of a traveller's diary, this book would not have come into existence. I cannot too cordially acknowledge my indebtedness to him.

I wish to acknowledge also my great obligation to Sir W. Garstin, who, on behalf of the Egyptian Government, has allowed me to reproduce the map of Lake Tsana, published in his Report on the Basin of the Upper Nile (1904); to Professor Poulton for his kindness in preparing the entomological appendix to this volume; and to Mr. C. E. Dupuis for permission to publish interesting photographs taken by him. And I have the pleasure of cordially thanking Mr. John Murray for leave given me to make extracts from "Life in Abyssinia," by Mansfield Parkyns, and the "Narrative of the British Mission to Theodore, King of Abyssinia," by

Hormuzd Rassam; Messrs. Macmillan and Co., for similar permission in the case of "The Nile Tributaries of Abyssinia," by Sir Samuel Baker; Messrs. C. Arthur Pearson, Ltd., in the case of "Abyssinia," by Herbert Vivian; Messrs. Chapman & Hall in the case of "A Narrative of a Journey through Abyssinia," by Henry Dufton; and Mr. Augustus Wylde in the case of "Modern Abyssinia." It would have been impossible to publish the collated information about the Soudan and Abyssinia contained in this volume without the privilege kindly granted me by these gentlemen.

A. J. HAYES.

SUEZ, 1905.

LIST OF ILLUSTRATIONS

MAPS

THE

SOURCE OF THE BLUE NILE

CHAPTER I

IN October, 1902, I was acting as Medical Inspector of the native quarter of Alexandria. Cholera had kept the staff busy, and we had brought the number of cases down to about fifteen in the last days of the month. I had passed the disinfecting gangs in review one morning before sending them to their allotted quarters of the town, and had just resumed my work in the hospital, when I heard a voice sing out to me—without preliminary—"Hayes, would you like to go to Abyssinia?"

It was the Chief Inspector of the Sanitary Department who asked the question. By the evening I had made up my mind, and told him that I should be glad to take the chance. However, I heard nothing more of the matter till the middle of November. Then I received a telegram directing me to make my preparations in Cairo. And I had no time to waste; for I learned that the expedition to which I was attached would start in seven days. The "jumping-off place" would be close to Khartoum.

Twenty years earlier the journey thither would have had a lingering fascination of romance. Ten years earlier it would probably have brought the traveller by painful paths to the Land of No Return. To-day it lies within the range of the prudent tourist who prefers safe adventures between regular meals. He is much indebted to Lord Kitchener.

In Cairo I joined my companions Mr. Dupuis and Mr. Crawley. They had had arrangements for the expedition in hand for many months; indeed, it had been planned for the previous year, but the illness of Sir William Garstin at Khartoum delayed it. On November 27 we reached Assouan. Like hundreds of my fellow-countrymen, I photographed the great dam, and the temple of Philæ. Our next stage was by post-boat to Wady Halfa, where we arrived on December 1. In the evening of that day we took train on the Soudan Government Railway. The following morning I had a new experience; for I had never before seen the mirage from the window of a railway carriage.

The granite rocks seemed to rise out of a great lake. The illusion was so perfect that I used it as a test of my black "boy's" shrewdness. I called him, and said, "Look at the water!" His answer was, "It is not the water. It is the water-thief, your excellency."

In this desolate land, with a burning sky over it, there are no trees. It is all of a piece, and the different parts of it give no sense of varied locality. The stations are not named but numbered, and here the water is stored in zinc tanks. Command of the tanks means command of the country—an important fact in connection with possible *émeutes* of native troops.

At Abou Hamed we reached the Nile again, and saw vegetation. Warm baths are provided in the station. We jumped out of the train in our pyjamas and rushed into luxury.

The railway follows the river to Helfiah, on the

opposite side of the river to Khartoum. Here I first beheld our camels, the steadily stubborn source of infinite vexation. All that is hard and heart-breaking in the character of the desert is incarnate in the camel. At Helfiah we engaged Soudanese "boys," settled matters with officials, and finished our preparations for shooting big game.

I went to the Soudan Club at Khartoum, and strolled about the town in the spirit of an inquiring tourist. It is carefully administered, and is laid out in three sections. The best houses are being built by the river-bank. Plans of these must be submitted for approval. They are mostly of red brick, and well constructed. Smaller houses, the "second class," have their sites behind the big residences. The "third-class" district is still further back. The object of this caste system applied to builders' enterprise is to secure that there shall be a fine promenade along the river front, and no mean streets near to it. The British and Egyptian flags are hoisted side by side over the quarters of the army of occupation. The Gordon College—the other tribute to the man's work—is a big, square, red-brick building. It seems to insist on its resemblance to barracks. No dahabeahs are seen on the Nile here; the boats are feluccas. Having made these notes of a tripper I returned to Helfiah. This was on December 5.

We dined in our camp, and slept in our tents for the first time, and I ceased to be a tourist, and became a member of an expedition. My companions and I showed the servants how to pitch the tents and peg them, and we apportioned loads to the camels. The camel is—not metaphorically—foul-mouthed. We put some of the boys up to try the paces of the animals, trusting chiefly to one who had served in the Camel

Corps under Kitchener. I watched the exhibition of ungainly shuffling with the face of one who judges, but I thought of the hours during which I should be shaken under the glaring sun.

Next morning we were up at sunrise. We had not yet lost touch of the British breakfast-table, for we ate sausages and poached eggs. Then the camels were loaded and the tents were struck, and at mid-day we started—three Englishmen, six native servants, forty camels, and ten camel-men. Our route lay along the north bank of the Blue Nile, over a beaten track in a wilderness of loose sand. A few scattered thorny mimosa bushes are the only vegetation. I timed the camels and found they travelled about $3\frac{1}{4}$ miles an hour. At two o'clock the temperature was 104° F. in the shade, but all the surroundings were new to me, and I enjoyed the journey in spite of the monotony of the country. It is all bare desert, and one never loses sight of the mirage. At one point I watched a flock of goats, which looked exactly as if they were knee deep in water. We reached Soba, fourteen miles up the river, at sundown. Here there is a pleasant and clean little rest-house built of straw. We had kept ahead of our baggage-train, which arrived an hour later. People in England who speak of tea as refreshing do not know the full meaning of that word. The cups were filled for men who had fasted since breakfast at 7.30, except for a welcome drink of river water on arrival, and were served with biscuits and sardines. After this, the happiness of a wash, and an eight o'clock dinner. A meal of curry and rice and tinned roast beef is not a severe hardship, and the only *contretemps* on our first day in the desert was the discovery that we must put ourselves on short rations of candles. A

THE START AHEAD OF THE CARAVAN. [*See p.* 4.

CHILDREN AT NO. 6 STATION, BETWEEN HALFA AND ABOU HAMED, EAT THE REMAINS OF LUNCH, CURRY AND RICE. [*See p.* 2.

mistake had been made about the quantity ordered, and we were ill supplied. But there was a clear moon. We turned in to sleep in the open, and the sandflies found us and were glad; for we were fresh blood, and they were small enough to crawl through the meshes of the mosquito curtains.

The first signs of dawn brought us out of bed next morning. We were in good luck; for the servants had kept a fire alight all night, and hot water was ready for us. There was beatitude in a warm bath after the plague of sandflies. We started at daybreak; there was no change in the track or the country. Heat, a high wind and dust spoiled the journey in the morning. We halted to lunch under the largest mimosa bush we could find, which gave as much shade as a telegraph pole.

Crawley opened his big artist's umbrella, which should have been a successful aid to comfort, but it tried conclusions with the wind and was worsted. We started again at two o'clock, and arrived in the evening at our camping-ground opposite the pleasant village of Sheik-el-Obeid. Our guns played the part of poulterer for us and varied our diet. In the morning Dupuis had shot a brace of sand-grouse, and in the khor * below our camp I had the luck to bag a teal. The temperature fell rapidly after sundown, and the night was very cold. The mosquitoes were numbed and did not stir, but the camels did, and I found them as effective in promoting a vigil. All night long they were browsing on the mimosa scrub; they relish the thorns, which are about two inches long. I heard them pull the branches, and then, when one was released, it recoiled into place

* A khor is a gully or gorge formed by the rush of rain in the wet season. It is a watercourse at that time, but a dry ravine during the rest of the year.

with a swish. There was a continuous noise of this movement, and it vaguely reminded me of breaking water.

The camels were hobbled, and stumbled from bush to bush. I constantly expected that one would stumble upon my bed. They loomed up, shadowy and grotesque, in the light of the bearers' fires, and I had a kind of nightmare of them without sleep.

On the 8th we were up before sunrise. Probably the camels, after supping all night, were tired. I say "probably"; for the camel is an aberrant type, and one can only guess its sensations. They were unwilling to barak * to receive their loads, and were grunting, snarling, and growling. When one of these beasts seems likely to become unmanageable, the camel-man "bridles" it by passing his index finger through the orifice between its nostrils. In breaking in young ones, a stout thorn, shaped as a peg, is used in the same position.

We started at 7.30, and walked ahead of the baggage train with our guns. At nine o'clock, when the sun had gained power, we took to the saddle. The country through which we passed is cultivated during the rainy season; it is scored with ravines, hollowed out by the rush of water in the wet months, and is uncomfortable ground for travellers.

During the day we shot five pigeons and a sand-grouse. Beside the river we saw and heard great numbers of demoiselle cranes, but found it impossible to get within gun-range of them. They have scouts on the watch both when they are in flight and when they settle, and as soon as the warning cry is given by the outpost the shooter's chance is gone.

* Kneel.

We reached our camping-ground, near Bushagra, before five o'clock, and pitched one tent for our own use in case a dust storm should come upon us. I bathed on the brink of the river, but kept out of the pool for fear of crocodiles. As yet I had seen none, but I made their acquaintance quite closely enough later in the journey. The night was warm, and we slept in the open on the bank of the Nile.

I began the next day (December 9) with a lucky shot. I had put my gun and a No. 3 cartridge—the only one I had—on the table near my bed before turning in, as I hoped that the demoiselles might be less cautious before daybreak. I was up an hour before dawn, and loaded my gun. I could hear the croaking of the birds for a long distance, and knew from the sound that they were moving. Just at the moment when I was ready, three cranes sailed over the camp. They were within range; and when I fired, a fine demoiselle collapsed and came to the ground in a heap. These birds make good eating.

All the country in this district has the same character. The low mimosa scrub is varied by patches of land which are cultivated in the rainy season. The Khalifa had his powder and cartridge factory at Bushagra, and possibly the natives thought that European vengeance might fall on them for this reason. In any case the Soudanese here showed nervousness, and made off when they saw us coming.

We ate another British breakfast—porridge and sausages—and then started on foot with our guns at half-past seven. A tramp of five miles gave us no sport, and we mounted our camels about nine o'clock. Our road lay through the hottest country I have ever known, and the temperature rose hourly as the day advanced. Smoked glasses ease the eyes a little, and there is shade under one's white umbrella, but nothing seems able to allay the thirst which this land causes. It is a mistake to suppose that one must go east of Suez to gain experience of true human drought. The water in my bottle * lasted, and only just lasted, through the journey.

Hot, parched men, in a hot, uninteresting land, are not likely to be in their best temper, and we had to bear with our guide. He had asserted that he knew the whole country, but he barely knew how to follow the track, and seemed to lack ideas of time and distance. We asked him, "When shall we reach the river again?" He pointed to a large section of the heavens and answered, "When the sun reaches there." He could not tell us the names of villages which lay on our route, and our maps were not trustworthy.

We reached the river bank after our halt at lunch-time, but left it again, and my camel's action appeared to be more back-breaking and monotonous than ever. Two hours in the saddle seemed like four, and Xenophon's Greeks were not more delighted to behold the sea than I was by the sight of the blue water as we approached it late in the afternoon. Our camping ground was beside the stream, and I rushed to the brink with my enamelled mug; the water was rather muddy, but the draught seemed like nectar.

I tried for a fish, without success. When I returned to camp I heard a camel grunting loudly and dismally. On going to discover the cause, I found the animal fastened to the ground by cords, while a man was scarifying its

* These bottles were of aluminium covered with felt. Before we started on the day's march, the felt was soaked in water, and the evaporation from it tended to keep the drinking supply cool.

back with live cinders. The object is to produce counter-irritation as a remedy for cysts. I arrived too late to deliver the unhappy beast from the horrors of Soudanese veterinary treatment, as the operation had just been completed. However, I gave the Arabs "a piece of my mind." They may not have understood my Arabic, but they could not mistake the meaning.

We had covered twenty miles during the day and camped beside a khor, opposite to a village called Baranku, which is situated on an island in the river.

On the following morning we started with our guns and walked ahead of the camels. Besides pigeons, we shot plovers of two kinds which I had not seen before. Our route throughout the day was over fertile soil, perfectly adapted for the cultivation of cotton. At present, grain (durrha) is grown here, and apparently only one crop is raised annually, after the summer rains. But Dupuis told me that if a canal were cut and a system of irrigation established the soil would bear "the kindly fruits of the earth" all the year round. Villages are numerous in this district, and we never lost sight of one or other of them during our day's journey. I noticed a great disproportion between the numbers of male and female inhabitants. In a hamlet in which I made inquiry there were eighteen women to one man. Nearly all the men had perished in the days of the Mahdi and Osman Digna. As a consequence, the women worked in the fields, while their few surviving consorts led lives of idleness. I suppose they were considered, and considered themselves, too valuable to be subjected to fatigue. They were too poor to buy tobacco, or smoking might have kept them silent. As it was, when they were not asleep they were talking, and I wondered how they could possibly find topics for incessant conversation.

We halted for the night on the bank of the river, two miles from Rufa'a, which is the second largest town on the Blue Nile. Miralai* Blewitt Bey, the Moudir† of Rufa'a, sent a police officer with a courteous message, expressing his desire to do all in his power to assist us. A welcome gift of eggs and fresh milk reached us a few hours later, and at the same time two night-watchmen (Gaffirs) arrived for the protection of the camp. We were heartily grateful for our countryman's good-will. Dupuis dined with him that evening, while Crawley and I remained in charge of the camp.

In the course of the afternoon I applied iodoform to the back of one of the camels, to keep the flies from a sore. All the drivers came clamouring to me after this, importuning me to treat every scratch on the beasts of burden in the same way. One wishes to conform to the tradition of ready helpfulness which is, happily, associated with the medical profession, but I declined to become physician in ordinary to the forty camels, as far as abrasions were concerned.

Next morning our troublesome guide sought me out to complain of symptoms of dyspepsia. He remarked that he was incessantly spitting—a circumstance which I had already noticed with disgust. I told him that if he swallowed the water which Allah gave him in his mouth, it would make his food move from the place where he had the pain, which was a punishment from Providence for his wastefulness. He asked many questions to make sure that I knew what I was talking about, and then left me, much impressed. My "prescription" had the desired effect, and he gave up his objectionable habit.

The morning of December 11 was very cold for

* Lieutenant-Colonel. † Governor.

these low latitudes. Just before sunrise the thermometer sank to 43° F. When we were ready to start, we found that the head baggage-man was drunk. He had been tippling "marissa," which the villagers brew from fermented barley. The man leered vacantly and was incapable of work. Considering the thirst and the opportunities there was little drunkenness among our boys in the Soudan, and we had no further trouble from this source. But the precepts of the Prophet did not keep those who accompanied us into Abyssinia from the vile beer of the country.

Blewitt Bey, on his pony, met us just after we had mounted our camels, and rode five miles with us. He gave us most valuable information about the land in which our road lay, and we were sorry when he said good-bye with best of wishes for our success.

We were in the saddle nearly eight hours and travelled about twenty-three miles. We reached our camping-ground at Abou Harras at five o'clock. The sun set almost immediately, and we, who were in advance, pitched our three tents by moonlight. The last of the baggage train did not arrive till an hour and a half later, and the drunken "Sheik of the Afsh" * was not with it.

The country through which we passed was similar to that which we had seen on the previous day. It is perfectly fertile. We had left the sand and mimosa scrub behind us, and were in a well-wooded region, with abundant undergrowth of bushes. The open ground is covered by grass which grows to a height of nine feet in the rainy season, but was now dried and matted under foot. It supplies pasturage for cattle, and I saw a herd of kine grazing on it. They were in fine

* Chief baggage-man.

condition. I handed my mug to the herdsman and asked him for some milk. He gave it to me willingly, and I thought it the best I had tasted since I left England in 1899. It seemed to me that this district offered a valuable field to British capital and enterprise.

I extract from my diary the following unsystematic notes, with an apology to the reader:—The butterflies which I saw most frequently on the journey were the clouded yellow and the red tip. The brimstone and painted yellow are seen occasionally. The familiar cabbage white is found on the banks of the river. I heard no singing bird except the chaffinch, but I saw a pair of blackbirds, tits of many kinds, and fly-catchers. A long-tailed species, which is a little smaller than a ring-dove, is very common. It was called "albicora" by our black boys.

THE WELCOME AT ABOU HARRAS.

[See p. 13.

CHAPTER II

We did not expect to receive a public welcome at Abou Harras, and were thoroughly surprised when the entire village, on the morning of the 12th, turned out to give us a reception. Women and men bore down on us, dancing and clapping their hands to a hideous accompaniment of tomtoms and drums. Many of them leaped and capered in the craziest fashion to emphasize the display of pleasure at our arrival. The women and children expressed their delight by a peculiar tremolo cry, which reminded me of that heard in Lower Egypt at marriage festivals.* One old man came on at a leisurely pace giving an exhibition of sword-play in our honour. He afterwards delivered a panegyric of the English, declaring that they were the saviours of his country, that the children of the Soudanese were now no longer torn from them, and so forth. I found that the young men were almost all absent from the village, doing military service on the Abyssinian frontier; the women who came out to show their delight officially by their shrilling were "the girls they left behind them." This singular deputation

* This trill is maintained on the same note—the upper B flat of the feminine vocal compass—and is produced by a vibration of the tip of the tongue on the hard palate. Many European ladies have tried to give a rendering of it, but without success. In Lower Egypt the "joy-trilling" is called *zungareet*.

had not approached very near, and presently retired a little way. Some of the slower movements of those who took part in the ceremony were by no means ungraceful, but the din was indescribably discordant. I went to inspect the procession at close quarters, supposing that it would shortly withdraw and disperse. But the "music" continued for fully an hour and a half, and it was more difficult to endure patiently during the latter part of the time as we were all busily writing letters for the homeward mail. A messenger had to carry them half a dozen miles and return immediately to rejoin, and he was waiting for our correspondence.

The point of junction of the river Rahad with the Blue Nile is half a mile distant from Abou Harras, and my companions walked to the spot later in the day. I attended to some odd jobs of the kind which a traveller always finds he has in arrear, and then I tried for fish, but caught nothing and lost my spinner. Many herons had taken their station on a strip of sand about a hundred yards from the place where I stood, and paid no heed to my presence.

On this day we had the good luck to receive two English visitors. The first was Mr. Wilson, the Moudir of Wad-el-Medani. We could only offer him a cup of rather muddy tea; but the village offered him the public reception, and the music was as loud as ever at the second performance. Our second guest was Bimbashi Gwynn, who landed from the sternwheeler which had left Khartoum on the 10th. He was on the way to Abyssinia to settle the boundary question. He had been in that country before, and as he was able to stay and dine with us, we had the benefit of many useful hints in the course of a pleasant chat in the evening.

VILLAGE MUSICIANS AT ABOU HARRAS. [*See p.* 14.

At Abou Harras we parted company with our guide, drivers, and forty camels. There was no unhappiness at the leave-taking. We inspected the men and animals that we had hired in their place, and devoutly hoped we had made a better choice.

On the following morning the thermometer again sank to 43 degrees, and the atmosphere and the water seemed bitterly cold to us. We started on the road to Gedaref just before nine o'clock. Our new guide was a Greek, who had made the journey many times, and was said to know everything that need be known. I was sceptical.

Our route followed the course of the telegraph, and lay at first beside the Rahad. This river differs considerably in the dry season from the Blue Nile. In the latter we had found a fairly fast current flowing over a stony or sandy bed between banks composed of mud. The stream passed through a succession of shallows and deep pools. The Rahad, at the point of junction with the Nile, was entirely dry, and we proceeded some distance up the course before we found pools. The banks are high and steep.

I do not think that Sir Samuel Baker at all exaggerated the fertility and value of the land in this region. He wrote: "The entire country would be a mine of wealth were it planted with cotton." * The Rahad, as he said, "flows through rich alluvial soil; the country is a vast level plain, with so trifling a fall that the current of the river is gentle." † This circumstance would facilitate irrigation. Besides mimosa trees and much other timber, we found in this region matted growths of bushes and wide stretches of long

* "The Nile Tributaries of Abyssinia," 1867, p. 525.

† Ib. p. 523.

grass now tangled on the ground—all bearing testimony to the quality of the soil.

I soon perceived that my hope of a better mount was vain. My new camel was a failure from the start. His paces were indescribable, and when he trotted I felt like a bad sailor in a small boat on a lively sea. In an hour's time I had lost my temper and got rid of the camel. As the result, I was better served, and a beast was brought for me which shambled evenly, and allowed me to keep up with my companions.

We halted for lunch at a village called Hadeiba, where two sheikhs came out to receive us. They were very polite, but would neither eat nor smoke, as we were in the month of Ramadan. By-and-by they remembered that they ought to give us a present, and left us, saying that they would bring two sheep, according to the custom of the Soudan. Half an hour later we quitted the spot.

That night we camped at a place named Khor Abou Segeira, close to the Rahad. Near at hand was a muddy pool in the course of the river, and we had to draw our supply of liquor from it. As usual, we set our Berkfeld filters to work. A candle is a part of the apparatus through which the water filters, and we found that this incessantly became foul and required to be cleaned every three minutes. However, we replenished our stock of clear water sufficiently, and then went to bed by moonlight in the open.

We started at a quarter-past seven on the morning of the 14th. No tents had been pitched the night before, so none had to be struck, and our departure was earlier on this account. My companions and I took our guns into the bushy ground beside the river while the camels followed the track towards Gedaref. We were in search of guinea-fowl, which are found in great abundance on the banks of the Rahad. These birds run in flocks, and rise at about forty yards' distance when one tries to approach them.

I lost sight of my companions in the thick, tangled vegetation, and after waiting where I was for a while in case they should chance to rejoin me, roamed on. I found no guinea-fowl, but pigeons and ringdoves were in great abundance, and there were hoopoes, hornbills, kingfishers, and numbers of other birds. A little later I came to the banks of the river. Here I saw a company of grivet monkeys. These, according to Mansfield Parkyns, are called "tota" or "waag" in Abyssinia. He described the species as "a beautiful little greenish-grey monkey, with black face and white whiskers." * The natty little fellows whom I met by the Rahad just answered the description. They were clean and sleek, and looked as "spry" as cleverness could make them.

After following the course of the river for about half a mile I made towards the track and found it without difficulty, but I saw no sign of our baggage-train. However, one of our boys trotted up a few minutes later, and then I learned that the camels were some way behind. In fact, an hour and a half passed before they overtook me. When the others came up, I learned that they had supposed me to be lost and had scoured the country in all directions, and that the whole party had been detained till all the searchers were called in. I was intensely annoyed to find that I had been the cause of the needless delay.

We lunched beside another muddy pool in the course of the Rahad, and then moved forward to our camping-ground for the night, which was called Mesr-el-Ashir.

* "Life in Abyssinia," 1868, p. 101

During the journey we had noticed that our escort were driving two sheep. We made inquiry about them, and were told that they had been purchased by the Greek guide on the previous day. We had a very strong suspicion that they were the animals which the Sheikhs had promised us at Hadeiba, and that the subtle Hellene had benefited by the gift. One of the sheep was killed in the evening, and we received "a present" of a leg, some chops, and a couple of kidneys. The meat was excellent.

Our route now left the course of the Rahad, and we turned eastward towards Gedaref.

Our beds were spread in the open again. Orion was straight above me as I lay, before falling asleep, and Sirius nearer to the southern horizon. I looked for the constellation and the star each night when I turned in, and had a sense of being in the presence of old friends while I gazed at them.

On the following morning, December 15, I again changed my mount for the better, and rode a smoothly trotting camel at last. Our day's journey was a short one, and brought us to the last watering-place which a traveller on this road reaches before the wells of Fau. On the way my companions shot a couple of guinea-fowl, and we were glad to see them at dinner-time.

We camped at night close to a long range of hills of a granite formation called Gebal Arang. They rise to a height of three hundred or four hundred feet, and make a striking change in the landscape as one approaches them. I knew that our Berbereen boys had never seen such high ground, and watched them to observe what impression it made upon them. It made none. They are an apathetic race, philosophers of the *nil admirari* school by temperament.

THE WELLS AT THE FOOT OF GEBAL ARANG. [*See p.* 18.

On these inland marches we carried water in tanks, and were obliged to hoard it. We had to go back in the morning to the allowance which had served us for washing purposes overnight. In all this region, every condition of life, important and unimportant, depends upon the rivers. At Gebel Arang we saw gazelles and the antelopes called ariels for the first time. We also caught sight of a large bustard, and my companion stalked it unsuccessfully. It is hard work to carry a gun in a hot country where every bush, and even the grass, grows thorns that "mean business," and the man who fails is usually weary as well as disappointed.

Next day, December 16, we travelled at the foot of the hills. It was a pleasant journey, for there was verdure, and the tall mimosas threw shade on the track. Besides the flocks of guinea-fowl to which we were accustomed, we saw a hare—the first we had found in the Soudan. Before midday we reached the wells of Fau. The drinking-water here is of a fairly good quality, but not quite free from mud. Guinea-fowl again supplied our dinner. They are slow to rise, but not easy to shoot; for they run in the most nimble fashion among the thick covert, and to come within range, the "guns" have to run too. The sun is strong overhead, the thorns are strong all round, and the exercise is smart work, even for a man in sound condition.

In the evening a messenger from the Intelligence Department reached our camp. He brought us letters and papers, and I ascertained that he, on a fast camel and without baggage, had travelled in two and a half days a distance which we had covered in six.

Our halting-place was between two ranges of the hills. The wind blew hard during the night, and the current of it came racing down the valley where we lay. Great clouds of dust were carried with it, and the particles were driven thickly into every crevice and corner. Crawley and I tried to form a shelter by piling up our square boxes behind us, but this was of little avail. It was an excitement of a sort, but not pleasant to a sleepy man.

We lay later than usual next morning, December 17, as we did not intend to strike our camp till midday. My companions went to shoot the poultry, and I remained in my tent and nearly shot myself without expending a cartridge. Our plan for securing fizzing drinks was to put a sparklet in a bottle of water and screw in the top. I was busy upon this operation when the bottle which I was handling burst. As luck would have it, I received no injury, though I was peppered with fragments of glass. One particle entered the eye of my boy Achmet. I extracted it and found that no serious damage had been done to the organ. After this, we abandoned the use of sparklets, and fizzing drinks were things of the "dear, dead past," and the uncertain future.

We left the camp at midday, after six guinea-fowl had been brought in, and struck into the waterless country between our halting-place and Gedaref. The distance is seventy-two miles. We were obliged to carry the drinking supply for the whole journey. Our route lay along a good track that follows the line of the telegraph.

The soil in this region is soft, friable earth, perfectly adapted for cotton cultivation. But the district is unhealthy in the rainy season, when malaria prevails and the country is a reeking swamp of dead and living vegetation. It was perfectly dry when we traversed it, and we ran no risk of disease. The tangle of desiccated tall grass lay upon the ground, as in so many tracts of this part of the Soudan, and our road was bordered by mimosa shrubs. They grow thickly in places, but were not high enough to give protection from the sun.

We halted for the night at five o'clock, and unpacked only our beds and such things as were strictly necessary, for we were to start at half-past four next morning.

I was very sleepy when I was roused at 3.30 a.m. on the 18th. We had a cup of cocoa, and then left our camping-ground at the appointed time by moonlight. The air was chilly even after the sun rose at six o'clock, and the warmth of the day did not make itself felt before eight. At half-past ten we halted for a *déjeuner à la fourchette*, and rested for a couple of hours.

Darkness had fallen by the time we reached our camping-ground at the place called Terras Wad-el-Fau. We had covered thirty-two miles of our journey since the morning. At this spot we rejoiced to find that an Egyptian soldier in charge of two *fantasses* (water-tanks) had been sent to meet us by the Inspector of Gedaref. Our camel-men with the baggage were some distance behind us, and we lighted a bonfire of mimosa wood to cheer them forward. They overtook us at their leisure, and our dinner of stewed guinea-fowl was not served till a quarter to ten.

We were out of bed by five o'clock next morning, December 19, and hastened the preparations for the start. We had to cover thirty miles in order to reach Gedaref in the evening. All this country—I cannot call it desert, though the conditions are not dissimilar—is comfortless for travellers, but our last day's journey in it was the worst. The heat was intense; so much so that "resting" at lunch-time seemed less tolerable than movement on camel-back. To quench an incessant thirst we had only the water stored in our tanks, which was warm and far from colourless. Even the hint of shade given by the mimosa scrub had now disappeared, and our eyes were wearied by the monotonous stretches of land, in which nothing was seen but the dried and matted grass. It had been fired in places, accidentally I think; for though it is often kindled to clear the ground for crops, in this region there was no trace of agriculture. The flames had a ghostly, exhausted, unreal look in the burning sunshine. After a march that seemed interminable we beheld, from rising ground, three trees with verdure on them, and then we knew that we were near the boundary of this distressing district. A two miles' journey brought us to the house of Mr. Flemming at Gedaref *—a square building made of sun-dried mud. At our countryman's dwelling a warm welcome and cool drinks awaited us. Here we found Bimbashis Gwynn and Liddell. That evening Mr. Flemming gave us our Christmas dinner—and we reconciled ourselves to antedating it—with true British plum-pudding in the *menu*.

* "The great plain comes to an end within a few miles of Gedaref, the ground becomes uneven and rocky, and Gedaref itself is situated in an open valley surrounded by bare hills of basaltic rock."—Sir W. Garstin, "Report upon the Basin of the Upper Nile," 1904.

MR. FLEMMING'S HOUSE AT GEDAREF. [See p. 22.

REST-HOUSE BETWEEN GOZ REGEB AND ADARAMA. [See p. 214.

From a Photograph by Mr. C. E. Dupuis.

CHAPTER III

I HAD slept in the open by choice and felt it a luxury to awake on the following morning with no prospect of a fatiguing journey. At seven o'clock on December 20 we met again at Mr. Flemming's house, and busied ourselves with amateur photography.

"You are the last white men I shall see until June next year," said our host, "so I want to keep your faces."

After breakfast we proceeded to the selection of seventeen "boys," who were to go up with our donkeys from Gallabat to Abyssinia. The character of the country beyond the frontier of the Soudan renders transit by camel impossible. It struck me as an interesting fact, not without a bearing upon the doctrine of Evolution, that the change from a flat or undulating to a mountainous region is here accompanied not only by a variation in the means of access, in the climate, in the flora and fauna, but in the race of inhabitants and the creed which they profess. In fact, the highlands of Abyssinia may be said to constitute "an island on the land," if one may borrow the phrase which Darwin applied to the peaks of the Andes. And a similar peculiarity in the survival of types can be observed here as that to which the great biologist referred in the other connection.

Mansfield Parkyns pointed out that the Abyssinians

were, in his time, poor swordsmen, ridiculous in their practice of musketry and frequently wanting in genuine courage.* Their country has constantly been the scene of civil war and other dissensions. Indeed the Jews among them, to whom I shall refer in another chapter, at one time established a separate state in the "mountain fastnesses of Semien and Belusa, where, under their own kings and queens called Gideon and Judith, they

* "The shotel (sword) is an awkward-looking weapon. Some, if straight, would be nearly four feet long; they are two-edged, and curved to a semicircle, like a reaper's sickle. It is a very clumsy weapon to manage. Many of the swords are made of the soft iron of the country and bend on the least stress. The handles are made of the horn of the rhinoceros, sawn into three longitudinal pieces, and incised so as to end in sharp points parallel to the blades. The shank is usually clinched over a half-dollar beaten convex. I should scarcely mind a blow from a sword thus mounted, as, were the striker to give his wrist any play, in order to make his cut at all effective, he could not fail sending one of the highly ornamental but very useless points of his hilt into his own wrist. . . . In the use of the gun the natives are in general exceedingly clumsy. They prefer large, heavy matchlocks, to load which is a labour of some minutes. They carry their powder in hollow canes fitted into a leathern belt worn round the waist; and, having no fixed charge, pour out at hazard a small quantity into the hand. This they measure with the eye, occasionally putting back a little if it appear too much, or adding a little if it seem not enough. After this operation has been performed two or three times, till they are pretty well satisfied as to the quantity, it is poured into the gun-barrel. The proper charge is now tested by the insertion of the ramrod. Lastly, when all is settled, some rag and a small bar or ball of roughly wrought iron are rammed down. This last operation (with the exception that the ramrod often sticks in the rag for half an hour) is not difficult, as the ball is made of about a quarter of an inch less diameter than the bore of the piece for which it is intended. It is great fun to see these gunners, when taken unawares by a sudden alarm; one can't find his flint, another has lost his steel; then there is the striking of a light, blowing the match, priming the gun, fixing the match to a proper length and direction; and, lastly, sticking into the ground the rest, which nearly all of them use, especially if their piece be of the heavy description. There is one thing in their favour—that the mere sound of driving in the rest is generally sufficient to turn away the bravest Abyssinian cavalry that ever charged."—Mansfield Parkyns, pp. 236, 241.

maintained till the beginning of the seventeenth century a chequered and independent existence.*

In spite of these hindrances and deficiencies the Abyssinian people, government and creed, have survived among heights protected on one side by the sea and on the other by a tract of country which is uninhabitable and impassable from May to September, on account of the incessant torrential rains and the floods which they cause. If these favouring conditions had not existed, the Abyssinian nation would long ago have been "converted" to Islamism or exterminated, and in the former case they would have become assimilated by consanguinity to, if not merged in, the surrounding Arab and negroid tribes. As it is, the Abyssinians form, ethnologically, an entirely distinct race.

This digression has carried me a long way from the "donkey-boys" at Gedaref. They were paraded to the number of a hundred for our inspection—Dongolis, Abyssinians, Tokrooris, Dinkas, Hadendowas, men from every clan in the district. An uglier-looking crowd I never beheld, and, as I gazed at the sinister, villainous faces, I hoped rather than believed that our choice of seventeen would not include robbers or murderers. We adopted the plan of choosing one or two men from each of the different races. Experience showed that the policy was sound, for no tribal quarrel was raised during our journey in Abyssinia. None of the groups was strong enough to commence a feud with confidence.

The business of selection was completed after the haggling and confusion inseparable from a bargain among Orientals, and I spent most of the remainder

* "Wanderings among the Falashas in Abyssinia," by Rev. H. Stern, 1862.

of this and the following day in acquainting myself with the features of life in Gedaref. As this place is a type of the more important towns in the Soudan, in which Great Britain has a permanent interest, I hope the reader will tolerate the transcription of a few of my notes.

The inhabitants number about eight thousand, and live—to use Mr. Dufton's description—"in a number of scattered villages, the principal called Hellet-es-Sook, or the market town." * The dwellings in this part of the Soudan and in Abyssinia are of the same construction, and I cannot better convey a conception of their simplicity and their appearance to those of my readers who have not seen them than by quoting the following passage from Mr. Dufton's work:—"Some of the houses are built of sun-dried bricks and have flat roofs, but most of them" (almost all in Western Abyssinia) "are conical thatched huts. The latter are made in the following manner:—In the first place a circle of about twenty feet diameter is described on the ground and surrounded by strong posts, each a yard apart, which are interlaced with thin pliable branches of trees, the whole being covered outwardly with durrha stalks, tied together with the long grass common on the banks of the river. A roof is formed in skeleton on the ground. A number of beams, corresponding to that of the posts, are made to converge at the top, and are held in this position by concentric circles of plaited twigs; the whole being then raised to its position on the house, where it is fastened and made ready to receive the thatch of straw and grass. Not a nail or rivet of any kind is used in the construction of these buildings, and they are rendered totally impervious to the wet. Of course

* "Narrative of a Journey through Abyssinia," by Henry Dufton, 1867.

their disadvantage in comparison with mud houses is their liability to catch fire, one or two cases of which I have witnessed; but they are generally built at sufficient distance not to endanger others, and a house of this description is rebuilt in a day or two." * Hence the scattered appearance of the huts. They cover a distance some five miles in length at Gedaref. These dwellings are called "tokhuls," and I shall have occasion to allude to them frequently under that name. In the Soudan storks build their nests round the upper part of the roofs, and are not molested.

In Mr. Dufton's time an open-air market was held twice a week in Gedaref. He thus describes it:—"One part is devoted to the sale of camels, cattle, sheep, and goats; another to that of corn and durrha; another milk and butter; another raw cotton, dates, etc. The butcher slaughters his meat on the spot, whether ox, sheep, or camel. The camel is killed somewhat differently from the rest; for it may well be imagined that a camel with merely its throat cut might be a long while in dying. They therefore adopt the mode of striking it with a long knife at once to the heart, having previously taken care to tie his long legs in a manner that he cannot move." †

The commerce of the place has been resuscitated since the British and Egyptian occupation. The market is now held daily. In addition there are the usual bazaars of an Oriental town. Gedaref has the advantage of possessing wells from which water of good quality is drawn. The place suffered severely in the time of the Dervishes. They raided the district as often as their supplies ran low, ruined the cultivation, and destroyed the greater part of the habitations.

* Dufton, pp. 16, 17. † Ib. pp. 35, 36.

The people seemed to be peaceable and contented under the present administration, and I was struck by their politeness towards Europeans. As we passed, they rose, if they were squatting on the ground; those riding dismounted as a token of respect; those who were at work discontinued it and stood still.

There are a certain number of shops in the town kept by Greeks. The commerce of the Eastern Soudan is almost entirely in the hands of Hellenes, who import goods by way of Suakim. The merchandise is brought up the country by camel-train. The articles most in request among the natives are cheap mirrors, tumblers, coloured silk handkerchiefs and similar goods—in short, cheap finery and cheap small commodities. I am not a commercial man, and do not speak upon the point with certainty, but I think German houses meet this demand. The Greek retailers also carry on a brisk trade in olives, which are imported in barrels. Olives and bread are the staple of the diet of many Soudanese, and they eat this simple food with relish.

We found that we could purchase sugar, taken from the familiar sugar-loaf, soap, similar sundries and tinned provisions at Gedaref. I was pleased to observe —for my own sake and that of British enterprise—that the latter bore the trade-mark of Messrs. Crosse and Blackwell.

My impression with regard to commercial opportunities in these towns is that a northern-bred storekeeper could not compete with the Greek retailer. I do not refer, in making this inference, especially to the singular astuteness of the latter. But it is an important consideration that these men are able to live almost on the same scale as the natives, and their profits are due to the frugality of their life.

We managed to start our baggage-train on the journey from Gedaref to Gallabat by a quarter to eight on the morning of the 22nd December. Then we breakfasted with Mr. Flemming, and mounted our camels to take the road again at 9.15. We moved forward at a trot, and I, now making trial of my fourth "steed," found I had good luck for the second time. But I do not think that any one could devote to the camel the affection which is so readily given to a horse after a brief acquaintance. The animal with the "sculptured sneer" does not invite friendship. It grunts and grimaces when one mounts, and the rider can be under no illusion as to its sentiment towards him.

We saw some ariel, and my companions gave chase and tried to get within rifle-range. But the animals were shy and wary. The hunters exhausted their supply of ammunition, but brought no game. We lunched as usual in the scrappy shade of a mimosa bush, and reached Shisana well shortly after four o'clock.

We had now left the "cotton soil" region, and entered an undulating country, where our road lay over rocky or stony ground. The character of the vegetation changed, and we began to pass through glades of the great mimosa forest which extends from the neighbourhood of Gedaref almost to Gallabat. Here the mimosas are not bushes but trees, and it is from them that the gum-arabic is collected which gives Gedaref its highly valuable trade and its prosperity. Large quantities of the gum are exported by way of Suakim, and there is a market for it at Cairo and other towns of Lower Egypt. For instance, much crape is manufactured in Damietta, where I have watched the process. Gum from Gedaref is used to stiffen the texture, and serves the same purpose in the preparation of various silken fabrics.

The industry might be largely developed. At present it is in the hands of Greek traders. As the supply of coin in the Soudan is small, the gum is used as a kind of currency by the natives, who barter it for other goods. A small duty is levied by the Government on the quantity exported, and no one seems appreciably the worse. I present this detail of information to fiscal controversialists, and make no demand upon their gratitude in doing so.

The gum-bearing mimosas are, on an average, I should say, about twenty-five feet high. The boles are straight. Thorns grow thickly on the branches. The white bark gives a characteristic and almost haunting feature to the great forest. The gum exudes, chiefly, at the junction of the branches with the trunk. Its appearance has been excellently described by Sir Samuel Baker: "At this season the gum was in perfection, and the finest quality was now before us in beautiful amber-coloured masses upon the stems and branches, varying from the size of a nutmeg to that of an orange. . . . This gum, although as hard as ice on the exterior, was limpid in the centre, resembling melted amber, and as clear as though refined by some artificial process. . . . The beautiful balls of frosted yellow gum recalled the idea of the precious jewels upon the trees in the garden of the wonderful lamp of the 'Arabian Nights.' The gum was exceedingly sweet and pleasant to the taste."* The trees were in flower when we saw them, and the blossoms scented the air sweetly.

The present method of collecting gum is primitive and unsystematic. Natives go a two or three days' journey from Gedaref into the forest, when they are pressed by want of a commodity to sell or exchange,

* "The Nile Tributaries of Abyssinia," 1867, pp. 356–357.

GUM BEARING MIMOSA TREES. [*See p.* 30.

TRYING THE TEMPERATURE OF A PATIENT AT THE DOCTOR'S PARADE.
[*See p.* 115.

and return with as much as they have gathered at hazard.

Throughout the day's march we had seen fires raging in the dried grass, and we had frequently passed places where the charred surface showed the effect of the flames. The custom of setting the undergrowth alight would destroy the mimosa forest, but for the fact that the fires burn low and travel quickly. The branches of the trees do not spring from the lower expanse of the boles, and there is consequently no combustion of the timber, except where a quantity of old, sapless wood is ignited.

Trouble threatened us at Shisana well. Grain for the donkeys and boys, that were to accompany us into Abyssinia, had been packed in sacks at Gedaref. The friction caused by the camels' movement had frayed these open, and the grain was lying on the ground when we arrived. At first we feared that we should be obliged to return to the town and purchase new sacks; but luckily the camel-men were bringing with them a stock of capacious and pliable baskets, made of slit palm-leaves, which they had intended to sell at Gallabat. These we commandeered, and so extricated ourselves from the difficulty.

Our journey next day was through entirely similar scenery, over undulating, rocky ground. We saw some ariel, but again they were too shy for us. Then we "declined upon" guinea-fowl, and bagged two after a brisk, exciting run.

The heat was overpowering, and at lunch-time we had no rest. Bees abound in this region, and get their honey from the flowers of the mimosa trees. As the land is almost waterless in the dry season, these insects suffer much from thirst. Even the moisture of perspiration attracts them in their parched state, and, in addition, there was the smell of water from our drinking-supply. The result was that they swarmed upon us; in fact, they mobbed us. Every drinking vessel was crowded with them. Our boys drank from calabashes; when these were put upon the ground, bees clustered on the edges and crawled towards the liquor. Impatient successors thronged upon the first comers and pushed them into the water, so that in a few minutes the surface was a mass of "struggle-for-lifers." In spite of the heat we had to keep moving; for when we settled, so did the bees—all over us.

That night we pitched our camp in the mimosa forest, six miles from the village and watering-place called Doka.

Next day—Christmas Eve—we started at seven, carrying our guns, and had not walked a mile when we disturbed a flock of guinea-fowl. Away they sped through the tangle of dry grass, and we after them. We blazed away here and there, but the birds, hit or not, were out of sight in an instant amidst that cover, and we missed in this way many that we had not missed in the other. Our bag consisted of four. When a bird dropped, the Soudanese "gillies," who were carrying our rifles in case big game should be seen, pounced upon it and decapitated it. Otherwise it would not have been clean meat for a Mohammedan. The land around Doka is hilly, and the ground rises to a height of which Primrose Hill would be a good example. We reached the village about nine o'clock.

Here we found, to our surprise, that the camel-men wanted to water every beast in our convoy. We were but a few miles from another watering-place, where the business could have been done in the evening without wasting the hours of daylight in which we were able to travel. We compromised by hurrying up the animals as soon as we had filled our tanks, and when half a score had drunk, the well ran dry.

We pushed on through the mimosa country. The temperature was about 102°, and I found the heat extraordinarily oppressive. The trees around us and overhead—they grow thickly in parts of the forest—seemed to shut out every movement of air, such as had always relieved us a little in the open tracts. Our extreme thirst constantly tempted us. We had made it a rule not to draw upon our ration of water till eleven in the morning, and abstinence till then was Spartan discipline. If one began to drink, it was a sheer impossibility to leave off, and the supply did not hold out. On calm reflection at a distance, I doubt whether our normal breakfast of tinned sausages and tea was wisely selected under the circumstances. But it could be quickly prepared, and we seldom had time to be leisurely.

While we were traversing the forest, I neither saw nor heard any bird except the guinea-fowl. The dried grass in which they wander was the only undergrowth. Usually the place was as still as a tomb. Of larger animals I saw none but antelopes, and not many of these.

We halted for lunch in a small ravine, and the bees did not find us till we had nearly finished the meal. We "smoked" them with cigarettes, cigars, and a bonfire, to no purpose. Then we shifted our quarters, but they followed. All of us were stung, and we were not quit of them till we mounted our camels and outdistanced the swarms.

We camped for the night among the mimosas

about seven miles from the watering-place called Zeraf Zaid. I thought we had been delivered from the plague of insects, but I deceived myself. Our camp was stormed by hosts of small creatures—tiny beetles, flying ants, and the like. They thronged and crawled on one's candle, one's book, one's face, and one's hands, and I found it difficult to write my diary. Some of them—I do not know whether they were prompted by hunger or malignity—added insult to injury by biting us.

On the morning of Christmas Day we started with our guns in advance of the convoy and bagged two brace of the usual poultry. I had not hitherto found the country unhealthy, though the climate tries one's condition; but on this occasion I felt "out of sorts," and was glad—for a wonder—to mount my camel after the tramp in the sunshine.

At 9.30 we reached the village of Zeraf Zaid. Here a clean hut serves as a rest-house. We found the place pleasant and cool. Within it were trestle-beds, of the kind commonly used in the Soudan and called "angareebs." Sir Samuel Baker has given a concise description of them: "The angareebs, or native bedsteads, are simple frameworks upon legs, covered with a network of raw hide worked in a soft state, after which it hardens to the tightness of a drum when thoroughly dry. No bed is more comfortable for a warm climate than a native angareeb with a simple mat covering; it is beautifully elastic, and is always cool, as free ventilation is permitted from below." * I concur in this favourable opinion of the native bed, and hope that its value in cases of sickness or injury that are not serious may be brought to the notice of the medical

* "The Nile Tributaries of Abyssinia," p. 182.

authorities of the army. It would be easy to furnish tent-hospitals with this simple appliance in any part of the Soudan.

We rested in contentment, and then lunched while the camels were watered. Our convoy started at noon, and we two hours later. We overtook the baggage at half-past three.

At Zeraf Zaid there is an outpost of "the Arab Battalion," which consists of Soudanese natives and Soudanese (*i.e.* naturalized) Abyssinians,* Egyptian non-commissioned officers, and two Englishmen, first and second in command, with the rank respectively of colonel and major. The strength is about two hundred, and the permanent head-quarters are at Kassala. The duty of the battalion is to protect the frontier and suppress the slave raids which are frequently attempted by Abyssinian subjects. I doubt whether the latter part of the supervision duty is quite successfully performed owing to the smallness of the corps, and it is well to bear in mind that these raids may upon some occasion lead to serious political complications affecting the attitude of the Anglo-Egyptian Government towards the *Negus Negesti.*† He has but slight control over his more powerful feudatories near the frontier; moreover, the succession to the throne will assuredly be disputed upon the death of the present ruler, and the whole country will then be plunged in misrule and civil war. It is clear that a very serious state of affairs may at any time be established in this region. Questions arising from incursions made into

* For a variety of reasons—some of them creditable—a number of Abyssinians abandon their allegiance to Menelek and the Rases (chiefs) and become subject to Anglo-Egyptian jurisdiction.

† "King of Kings," title of the Abyssinian monarch.

territory where the British flag flies from districts where no effective or responsible government exists may assume the gravest importance.

The "Arab Battalion" is a sort of legacy to us from our predecessors, and the men still wear the picturesque uniform chosen for them in the days of the Italian occupation. It consists of a white tunic extending to the knees, loose white trousers, sandals surmounted by white gaiters, a broad green sash round the waist, and a red tarboush with a green tassel. I cannot offer a valid opinion upon the sufficiency of this body as a frontier guard in the event of disturbances over a wide area, and I am not able to judge what effect the racial sympathy of some members of it with Abyssinians "over the border" would have in such a case.

Shortly after we had left Zeraf Zaid on the road to Gallabat we saw a great bustard near at hand. One of my comrades tried to shoot it for our Christmas dinner, but his luck failed him. However, I had brought the materials of a plum pudding with me from Cairo, and hoped that it would be a solace and a pleasant surprise for the other two members of our mess. The cook obtained some eggs in the village, and I gave him careful and emphatic directions and swore him to secrecy. Ours is a phlegmatic race. The cook did his work creditably, the pudding came to table, and was eaten without comment. The poultry in these villages is of a stunted size, but the flesh is tender and makes good eating. Pigeons are to be had in all the hamlets.

Our camping-ground for the night was a pleasant spot, bare of grass, and surrounded by high trees. In honour of the day, and in expectation of entering Gallabat early on the following afternoon, I discharged a professional man's duty towards his toilet, and scraped a week's stubble from my face. In the wilderness even a doctor can neglect appearances.

On Boxing Day we entered rising country with features different from those seen in the mimosa forest. The sides of the rocky hills have been seamed with ravines by the rains. Nearly all the trees are still of the same order, but certain species have broader leaves, and their appearance resembles that of the timber found in the temperate zones. However, I saw no specimen of the varieties common in England. I heard one song-bird in this district, the chaffinch. Water-wagtails abound.

As usual, we made an excursion ahead of the baggage with guns and rifles. The game birds are sand-grouse and partridges, and we shot a brace of the latter. Crawley brought down an ariel buck, which was in good condition. This was our first venison. The shot caused intense excitement in our convoy, and the guides and boys all made a rush for the beast to cut its throat before it died of the bullet wound. It had been hit in the neck.

We rested in a ravine called Otruk—a lovely wooded gorge in which a small stream was flowing. A little further down the course the water disappeared in the sand. Here, for the first time, I saw a troop of baboons. No doubt they inhabit caves and clefts in the surrounding rocks. Some were fully of human stature. Partridges and guinea-fowl were numerous here, and beyond question the rivulet in the khor attracts animals of every kind existing in the district. It was the first running water we had seen since we left the Blue Nile.

We rested until two o'clock, and then rode on towards Gallabat.

When we were half an hour's journey distant from the town, we were met by Mr. Saville, the Inspector, who accompanied us to our quarters.

CHAPTER IV

Politically, Gallabat is a place of much importance, as the principal station on the frontier between Abyssinia and the Anglo-Egyptian dominions. It is the terminus of the telegraph line whose course we had followed along a part of our route, and I shall presently explain why I think it would be difficult to carry the wire from this point into Menelek's country.

The town consists of two main thoroughfares and of the scattered "tokhuls" which form the villages throughout Abyssinia and the Eastern Soudan. It contains about eight hundred inhabitants. Formerly, I have no doubt, there was a larger population, but this place, like all other trading centres in the land, suffered severely in the days of the Dervish ascendency. The "shops," which abut upon the main roads, are hovels in a row under a heavily thatched roof, and are separated by rough and ragged partitions of cane thatch. They are "business premises" only, and the proprietor and his family dwell elsewhere. A considerable trade is carried on by Abyssinians who come to Gallabat to sell coffee; sometimes they have a leopard's skin or other hunter's trophy to barter in addition. They exchange their goods chiefly for cotton, which is used in the making of *shamas* and other garments worn in Abyssinia. Sir S. Baker mentions a commerce in bees'-wax and hides, but I saw no evidence of it in Gallabat, and I think it must have declined.

The traffic in coffee proceeds regularly during the dry season, but ceases, of course, when the rains make the neighbouring Abyssinian hill-country impassable. A square, in which there are booths, is set apart for the use of the Abyssinian traders. In the middle of it are the large scales in which the coffee is weighed, and I was much interested to learn that not only the amount of the import duty payable to the Egyptian Customs, but also the sum charged as export duty by the Abyssinian Government, was here determined by the weight of the bundles. The latter tax is retained at the time in the hands of the Anglo-Egyptian officials, and is afterwards remitted by them to Menelek's Revenue Department. I need hardly add that this is a recent arrangement, and it seems to show that British influence in the Soudan has inspired confidence at Addis Abiba.

The town of Gallabat is five miles distant from the bed of the Atbara. It is built on the slope above a small watercourse. This rivulet and the adjacent supply of drinking water which the inhabitants use are —in all but details—as they were in Sir Samuel Baker's time. He thus described them:—"We were horribly disgusted at the appearance of the water. A trifling stream of about two inches in depth trickled over a bed of sand, shaded by a grove of trees. The putrefying bodies of about half a dozen monkeys, three or four camels, and the remains of a number of horses, lay in and about the margin of the water. Nevertheless, the natives had scraped small holes in the sand as filters, and thus they were satisfied with this poisonous fluid; in some of these holes the women were washing their filthy clothes." * There are three wells in the town from which good water is obtained. The soldiers of the garrison use this supply, and it is available for all who will take the trouble to draw from it. But generally, the inhabitants, either because they are too indolent to pull up buckets or from mere adherence to custom, prefer the "poisonous fluid" procured in the traditional way.

Our quarters were in the dismantled Dervish fort, which lies on the summit of the hill, above the town. The thick brick wall which surrounded it is now ruinous. In the enclosure are barracks for a detachment of the Arab Battalion, a small rest-house, and the telegraph station.

We set our boys to pitch three tents for us, and before long I had the indescribable delight of tubbing in my portable rubber bath.

After this we received, in the rest-house, the interpreter whom Menelek had sent to await us. His name was Johannes. I learned that he had been three weeks in Gallabat, so perhaps it had been expected that we should travel more rapidly than we did. With him were an assistant, by name Walda Mariam, officially attached to his "mission," and a servant.

Johannes was a tall, handsome man, with grizzled hair and beard. In person he seemed cleanly, but his *shama*—the toga-like robe, also known as "quarry," †

* "The Nile Tributaries of Abyssinia," p. 502.

† Mansfield Parkyns has given a full and careful description of the national costume. I extract the following details from it:—"The 'quarry' is the principal article of Abyssinian dress: it is of cotton, and very fine and soft; those of the richer being finer but probably not so serviceable as those of the poorer class. It is made in three pieces; each piece is about three feet broad by fifteen feet long. Near both ends of each piece is a red stripe, five or six inches broad." The pieces are sewn together so as to form "a white double cloth, with a red border near

which is universally worn by Abyssinians—was soiled. It is unfashionable, and even a breach of established custom, for a man to appear in a clean *shama* except at

the bottom only; the breadth of the 'quarry' is nine feet by seven and a half long. . . . The methods of putting on the cloth are as various as the modes of wearing a Highland plaid. One of the most ordinary ways is first to place it like a cloak over the shoulders: the right end, which is purposely left the longer, is then thrown over the left shoulder, and the bottom border, which would otherwise (from its length) trail on the ground, is gathered over the right shoulder" ("Life in Abyssinia," pp. 228, 229). "The trousers are of a soft textured but rather coarse cotton stuff, made in the country, and are of two sorts: one called 'cállis,' the other 'coumta.' The former reaches half-way down the calf of the leg, the latter to about three or four inches above the knee" (ib. pp. 225, 226). The other garment worn by the men is a belt. These cinctures "vary in length from fifteen to sixty yards, and are about one yard in width. In quantity of cotton they are nearly all of the same weight, as the very long ones are in proportion finer than the shorter" (ib. p. 227). With regard to feminine dress, "there is a distinguishing costume for young girls, and for those who, from being married or otherwise, are no longer considered as such. The dress of the former is indeed rather slight, though far more picturesque than that of the latter. . . . The girls merely wear a piece of cotton stuff wrapped round the waist and hanging down almost to the knee, and another (or the end of the former, if it be long enough) thrown over the left shoulder, so as to leave the right arm and breast exposed. In other parts of Tigre a black goat-skin, ornamented with cowries, is often substituted for this latter. An ordinary woman wears a large loose shirt down to the feet, with sleeves made tight towards the wrist. This, with a 'quarry' similar to those of the men, but worn rather differently, and a parasol when out of doors, is a complete suit" (ib. pp. 241–243). Parkyns thus describes the system of hairdressing: "In general, neither sex wears any covering on the head, preferring to tress and butter that with which nature has provided them. The hair of the Abyssinians is admirably adapted for this purpose, being neither short and crisp like a negro's, nor yet of the soft elasticity of a European's, but between the two. . . . The operation of tressing is a very tedious one, usually occupying an hour or two per head: therefore, of course, it is repeated as seldom as possible: by some great dandies once a fortnight: by others once a month, or even less frequently. In the interim large supplies of fresh butter are employed, when obtainable, in order to prevent the chance of a settlement of vermin; and a piece of stick, like a skewer, is used for scratching. The hair is gathered in plaits close over the whole surface of the head, the lines running fore and aft, and the ends hanging down in ringlets over the neck. . . . Some ladies have rare intervals. As all sorts and conditions of men in the country are infested by vermin it may be imagined that an Abyssinian's clothes usually add nothing to the pleasure of his company.

Our interpreter had spent three years in Marseilles, and we found that he and his followers conformed to the requirements of civilization as regards dirt much more closely than the generality of his countrymen. I do not know whether Parkyns's description of the use of butter as pomade still holds good in parts of Abyssinia; probably not very widely, as Mr. Wylde tells us that "European hats are getting very common, and are generally of the bowler, wideawake, or Terai patterns," * but in the western provinces, where nearly all the people whom we met were peasants, we saw no sign either of this innovation or that the dairy had been drawn upon to smarten the *coiffure*. Nevertheless, greasiness of person was the rule.

I may mention here that bodily cleanliness is not only unusual in Abyssinia, but raises a doubt as to the genuineness of a man's religious profession. In this respect matters have not changed since Parkyns's time. "St. John's is the only cleanly day in the calendar; for

their butter daubed on nicely, and then some scent: but the great 'go' among the dandies is to appear in the morning with a huge pat of butter (about two ounces) placed on the top of the head, which, as it gradually melts in the sun, runs over the hair and down the neck, over the forehead, and often into the eyes, thereby causing much smarting. This last ingression, however, the gentleman usually prevents by wiping his forehead frequently with his hand or the corner of his 'quarry.' As may be imagined, the dresses neither of the women nor men are long free from grease; but this, especially among the latter sex, is of no importance; indeed, many young men among the soldiery consider a clean cloth as 'slow,' and appropriate only for a townsman or a woman. These never have their quarries washed from one St. John's Day to another" (ib. pp. 243–245).

* "Modern Abyssinia," p. 248.

in the evening the whole population, male and female, old and young, go down to bathe. It is a fact, that, excepting on this occasion, there are many of the number, who, beyond washing their hands before and after meals, and their feet after a journey, never trouble the water from one year's end to another. My habit of washing every day in the European fashion gave rise to much scandal on my first arrival; and it was constantly inquired, 'Is he a Mussulman, that he thus washes, and so often?'" * The Abyssinians are fond of shaking hands, and it is impossible to avoid this civility without giving serious offence. Nearly all, owing to their independence of soap, suffer from a complaint once associated—wrongly, no doubt—with a highland region nearer home, and the result was that none of the Europeans of our party escaped *scabies*.

Johannes possessed a felt sombrero, which he held in his hand when we received him in the rest-house. Like most of his countrymen, he wore a cartridge-belt around his middle. He also carried a revolver in his belt. His "boy" held his rifle. Altogether the interpreter was a picturesque but piratical figure, and it occurred to me that a group of Abyssinian "villains" would be effective in a melodrama.

The interpreter bowed to the earth each time we addressed him. He had brought letters for us from Menelek and from Colonel Harrington, the British Minister at Addis Abiba, which is now the capital of the country. The king's letter is a kind of superior passport, which is generally granted to Europeans who enter his realm with his consent. It is useful, but is not too obsequiously respected in all parts of the land.

* "Life in Abyssinia," p. 282.

Johannes had a fair knowledge of French—a matter of importance to one of his calling, seeing that the Abyssinians and the French are neighbours, and that an increasing amount of the commerce and traffic of the country is likely to pass through Jibuti. A railway has been commenced from that point, and a concession obtained from Menelek for its extension into his dominions. Mr. Herbert Vivian has recently pointed out the important developments which may be brought about by the construction of the line.*

The Dervish fort at Gallabat is literally a "cockatrice den." We speared eight or nine fine specimens upon the walls, using our knives, and among them was the largest black scorpion I had seen. The rest-house swarmed with these disquieting arthrogastra, and one of our servants was stung during the evening. Parkyns remarked that, though he had several times endured the sting himself without serious consequences, he had heard of many instances which had ended fatally.† Our boy probably had in his mind a similar dismal record, for he howled till midnight. I had very little sympathy with him, especially as he prevented me from sleeping. He, like the other servants, had been provided with boots but persisted in running about the fort barefoot.

I was delighted by the change of scenery which commenced at Gallabat after the long tracts of desert and mimosa forest. To the eastward, the Abyssinian mountains were visible, and there is a most stirring splendour in the sight of the distant peaks when one beholds them in the pink, hot glow of sunset.

An escort of ten men had been appointed for us. All

* "Abyssinia," by Herbert Vivian, 1901, pp. 314–327.

† "Life in Abyssinia," p. 418.

these were Abyssinians by race, but "naturalized" Egyptians, serving in the Arab Battalion, and they spoke Arabic in addition to their native tongue. There was an odd significance in the fact that our journey to Lake Tsana familiarized the inhabitants of the western provinces of Menelek's kingdom with an Italian uniform which had become British.

On December 27 we "lazed" and enjoyed our indolence. Our only activity was to make trial of our patent donkey saddles. We had seventy of these, and found them anything but serviceable. The natives fasten the loads on the donkeys by long thongs of hide, giving about six girths for each animal. Our saddles allowed but one girth, and, as the consequence, the loads slipped round. On the 28th my companions went to take measurements in the course of the Atbara. I remained in Gallabat and superintended the conversion of camel loads into donkey loads. This involved getting the right weight of flour and corn—fifty pounds—into the sacks. It was troublesome work done by trying people, and the supervision of it would have taxed the patience of a saint.

On the 29th we were all busy with the baggage, sorting our supplies. We took with us sufficient for six weeks. The remainder was left in the zaptieh,* and we fully expected that the rats and the white ants would clear it up for us before our return. Our cook had a touch of malarial fever, and during the day I gave him thirty grains of phenacetine and nine grains of quinine. These quantities were effective.

We were to start on the following morning, and sat longer than usual over dinner, and many yarns were told about Egyptian officials. One—I hope it is not a

* Guard-house.

stale story to my readers—ran as follows :—An Egyptian battalion, under a native officer, had been sent to a remote district. The commander had orders to telegraph to head-quarters if anything unusual occurred. He sent off a message announcing that one of his soldiers had died suddenly, and that he awaited instructions. The reply was, "Bury the man, but make sure he is dead first." The native officer thereupon reported by telegram, "Have buried soldier, and know he is dead, because I hit him on the head with a fishplate." It need hardly be said that the fishplate was not of the kitchen but the railway kind.

On the following morning, December 30, we had further proof of the defects of the saddles which had been provided as baggage-gear for the donkeys. When, after much tedious work, we had balanced the load on an animal, we started it on the road and turned our attention to another. Five minutes later, the beast that had been despatched was brought back with its load under its stomach. My companions toiled on through the heat of the day; but I fear my services became rather like those of the fly on the wheel, for the malarial fever had got hold of me, and the commencement of a bout with it nearly took me off the list of effectives.

We made a start at last, a little after eleven o'clock, with the intention of travelling about eight miles, and crossed the stream below Gallabat by a stony ford, over which water a few inches deep was running. I was told that this rivulet is never quite dry, and I observed that in the bed of it were the stagnant pools usually found in the watercourses of this district. We had seen the last of the camels at Gallabat, and my companions and I now rode mules. They are the best mounts in the rough, steep, and almost trackless region which forms the western border of Abyssinia. Horses would be useless here for a journey, either to carry men or for purposes of haulage; and I doubt if ponies could anywhere be found that would be more than an encumbrance, at least for long and steady travelling at a fair pace. I believe that the Abyssinians of the Highlands use horses in their hunting expeditions and slave-raiding forays—at all events they do in their campaigns in their own country—and I have little doubt that the inhabitants of the lofty hill district adjoining the border tract, who are of mixed race and bandits by immemorial tradition, use any animals they can lay their hands on when they, like Roderick Dhu and his men, "with strong hands redeem their share" from travelling merchants or Soudanese villagers. But these are "spurts" of work, rapid dashes made by unencumbered men, not plodding regular marches, and, generally speaking, those who cannot pass through this country on a mule or an ass must use their own feet or not go at all.

When my companions and I had advanced about four miles, we halted for lunch at a pleasant spot beside a brook. After the meal, some guinea-fowl were seen, and one was bagged. No doubt many more were hit and lost, as usual, among the high grass. We were in the saddle again at three o'clock, and reached our camping-ground an hour and a half later.

The fever had a grip of me by this time, and I rested in bed, and soon felt the benefit of a dose of quinine. It enabled me to tackle my dinner, and that is no small boon to a sick traveller.

Our boys had cut down trees and formed a zareba, within which the donkeys were penned to prevent them from straying and give them protection from hyenas.* I lay idly watching the camp-fires and listening to the occasional outbreaks of scuffling and braying among the donkeys, and presently fell asleep till dawn.

The country which we entered on that day is full of rounded hills. Boulders crop out from the surface of them, and they are separated by ravines with sandy beds. There is no regularity in the configuration of these hills, and no uniformity in their size. When I saw them they were covered either with dried grass or with mimosa scrub.

The region which we had now reached is, as I have said, impassable in the wet season, and Abyssinia is then impregnable as far as its western boundary is concerned. The violence of the storms and the rush of rain are such that it is doubtful whether even a line of telegraph posts and wires could be carried through the district with a likelihood that it would be in working order for any length of time between May and September.

* "On one occasion we had a small adventure. We were resting one night near the summit of a mountain, when about two hours before daybreak we were awaked by a loud hubbub and the discharge of a gun. Starting to our feet, we inquired what was up, and our anxiety was increased by M. Lejean's Arab seizing the second gun and discharging it. All I saw, for it was pitch dark, was one of the mules kicking about amongst the ashes of a half-extinguished fire, and endeavouring to extricate himself from the leather thong which bound his head to a tree. This he soon succeeded in doing, and went off at a furious rate towards the woods which clothe the sides of the mountain. I thought he had burnt himself at the fire, and that this was the cause of his breaking loose, but the rest maintained that he had been bitten by the hyenas: and they were right, for at daybreak we saw and gave chase to one of these brutes, who was still prowling about. These creatures were much bolder here than near the villages. As we were sleeping in the open air, with our faces exposed, it was fortunate for us that the hyenas preferred trying mule's flesh to man's. We heard afterwards that the mule had returned to Gallabat, a distance of forty miles, but was so severely bitten in the flank that he was perfectly useless." (Dufton, "A Journey through Abyssinia," pp. 48, 49.)

The floods would probably be as prohibitive to an underground system of wires as to one overhead, and in both cases allowance would have to be made for the extreme electrical disturbances which often occur at brief intervals during the rainy season.

I have already referred to the raids that take place from the Abyssinian side when the country can be traversed, and will only add here that they are carried out frequently enough to maintain a supply of slaves for all persons in the country whose prosperity raises them above the degree of a peasant. For example, when we reached Gallabat on the return journey we found that eighty-three Soudanese had been carried off during an incursion which had been made since we left the place. It is needless to insist on the difficulties that may arise from raids of this kind undertaken by people who are, nominally at least, subjects of the Negus.

I can testify that slaves are found everywhere in Western Abyssinia, but among the female victims I never had the good fortune to behold one of the class which impressed Sir Samuel Baker so pleasantly at Gallabat. He wrote, "On my return to camp I visited the establishments of the various slave merchants; these were arranged under large tents formed of matting, and contained many young girls of extreme beauty, ranging from nine to seventeen years of age. These lovely captives, of a rich brown tint, with delicately formed features and eyes like those of the gazelle, were natives of the Galla, on the borders of Abyssinia, from which country they are brought by the Abyssinian traders to be sold for the Turkish harems. Although beautiful, these girls are useless for hard labour; they quickly fade away and die unless kindly treated. They are the Venuses of that country, and not only are their faces and figures perfection, but they become extremely attached to those who show them kindness. There is something peculiarly captivating in the natural grace and softness of these young beauties, whose hearts quickly respond to those warmer feelings of love that are seldom known among the sterner and coarser tribes. They are exceedingly proud and high-spirited, and are remarkably quick at learning. At Khartoum, several of the Europeans of high standing have married these charming ladies, who have invariably rewarded their husbands by great affection and devotion. The price of one of these beauties of nature at Gallabat was from twenty-five to forty dollars." *

* "The Nile Tributaries of Abyssinia," pp. 515, 516.

CHAPTER V

On the morning of December 31, the donkeys were laden and our journey was begun by a quarter to eight. We were again hindered on the march by the shifting of the donkeys' loads.

Our track now lay among craggy mountains, in the rain-scoured district, where there is no single human habitation. I saw neither wild beast nor bird, and the country gave a strange impression of stillness and lifelessness. The flora is quite distinct from that of the Soudan. The trees are tall and spreading and of many different species. I recognized the "Matabele apple," and saw the hard, uneatable fruit on the branches. The ground in many places was covered with "bamboo grass," too high for a mounted man to look over, and the thick, overtopping growth seemed to shut out the air. As we advanced during the afternoon we discovered that the grass had been fired at several spots along our route—I do not know by whom, or whether it was kindled accidentally or by design. The flames drove out some partridges, and Dupuis and I thought that we saw a tasty breakfast for the next morning. But when we moved forward with our guns the birds, to our amazement, ran back into the burning cover. The fire died out in the evening, and all was quiet at night.

There is much stony land in this district. The soil

of the fertile tracts is not of the friable kind which we had so often seen in the Soudan: but there are many fissures in it, due, no doubt, to alternate drenching and drying in the rainy season. Our road often lay along steep and wide gorges, through which an enormous volume of water descends to the Atbara at the time of the floods. My mule was very steady on his legs and the motion was easy when he trotted on open ground. He only showed the waywardness of his kind when I led him. It was pleasant to think that some sort of understanding existed between the rider and his mount after one had been accustomed for weeks to the impassive eye of the camel.

On January 1, a couple of shabby fellows visited our camp in the morning. We questioned them and found, from the answers which Johannes translated, that they were the escort sent to accompany Gwynn and were on the way to join him at Gallabat. He might consider their company a sign that he was honoured, but scarcely a sound protection against robbery. We struck our camp after this and commenced the day's journey, following the bed of a river. As usual, there were dry tracts and well filled pools in its course. Later, we climbed the bank and struck into the country beyond, continually riding up hill and down dale. Again we saw an abundance of tall grass and fine timber. The former was dry at this season, and had been burned in many places. The country here abounds in big game, but the height and thickness of the grass make it very difficult to sight animals. We saw a koodoo, three gazelles, traces of elephants, and spoor of other beasts, but none came within range.

We had covered fourteen miles by half-past eleven, and then halted, considering that the donkeys had journeyed far enough. Our progress depended on them, and we were careful not to overwork them. Three were suffering from girth sores. We pitched our camp in a ravine about a quarter of a mile distant from the bed of the Gundar Wahar—which is the river Atbara with an Abyssinian name.

One of our escort was reported ill with fever, and I found that his temperature was 104 degrees. But quinine had brought about a decisive improvement by nightfall. I had another patient. My friend Crawley had chafed his instep, and there was a bad sore on it. I had recourse to a little doctor's diplomacy to keep him still, and put on such a large fomentation that he could not walk. Then Dupuis and I took our guns and rifles and tramped in the jungle—for the mass of thick, lofty grass deserved the name. My companion shot a brace of partridges, and we saw some guinea-fowl, and much spoor of big game. But nothing more got into our bag, and the excursion was wearisome and disappointing.

On January 2 our road again lay through hilly, verdurous country. While the donkeys climbed or descended the steep inclines their loads slipped as before, and kept all hands busy. The boys were continually shouting for help, and the burdens were replaced amid yelling and cursing. These natives never worked without talking, singing, or swearing, and they were specially fond of hearing their own oaths. But, I think, very little ill-will went with the words, and in spite of the endless imprecations uttered over the donkeys, they treated the animals well.

During the morning I took my rifle, filled my pockets with ball cartridges, and rode ahead of our party in the hope of trying a shot at big game. But I found small parties of Habashes* at intervals along the track throughout a distance of two miles in front of our convoy. Clearly it would be useless to search for any large wild beast in the proximity of these groups. I recognized the men as folk who had attached themselves to our train. They made a practice of camping where we pitched our tents, and had hitherto forestalled us in the choice of ground, and settled under the most suitable trees. They sought protection from the robbers who infest the district—the most notorious at the time was a Soudanese Arab called Hakos. I was glad that these wayfarers should enjoy a sense of security, but resented having sport spoiled. So I addressed a remonstrance to Johannes, who promised that all camp followers should keep in the rear in future, and he was as good as his word.

That afternoon we pitched our tents on the banks of the Gundar Wahar, which was here a stream trickling from pool to pool. We had travelled about fourteen miles. My friend Dupuis fitted out his angling tackle, and tried his fortune in some of the pools, using an ordinary spoon-bait. He caught three fair-sized fish, belonging to the perch family. I saw him land the biggest, which weighed six pounds, and showed fight. It had to be played into

* "The word 'Habash' (the native name for Abyssinian) means, I believe, a 'mixture' in the Giz language—a mixture of various qualities of corn goes by that name in some of the provinces of Tigre. It is supposed by some that a great number of Jews followed the Queen of Sheba, on her return from her visit to Solomon, and that a large colony of fugitives also took refuge in Abyssinia about the time of the destruction of the temple and the captivity. Subsequently the Greeks sent missionaries, and they were doubtless accompanied by adventurers; and the Portuguese sent a number of troops, some of whom remained in the country for many years. The variety of complexion, observable in both sexes, is, I should think, attributable to the mixture of races of which the nation is composed." (Mansfield Parkyns, "Life in Abyssinia," p. 224.)

the shallows, and was brought ashore in smart style. Crawley, who was lame, limped to the edge of the river and cast a line with dough on the hook. He fished patiently, and his perseverance was rewarded with two little creatures of the size of sticklebacks.

All the camping-grounds by the waterside in this district are called *warshas*, and as there are few distinctive names for places in the uninhabited tract, we called the spots where we halted *warsha* number one, *warsha* number two, and so on.

In this region, near the rivers, the white ants are extremely numerous. They are not seen at a distance from water, and cannot work without moisture to renew the fluid that exudes between their mandibles. This enters into the composition of the stiffened earth of which they build their dwellings. They are voracious and destructive, and have a propensity for gnawing leather, felt—which unfortunately they found in the pads of the donkeys' saddles—and any textile material. They are most active at night, and, to preserve our bedding, we had provided ourselves with "Willesden sheets." These are made of canvas coated with a preparation of arsenic, which the white ants avoid. The sheets were always spread under our beds when we camped, and if anything that was to the taste of the ants slid beyond the edge in the dark, destruction awaited it.

I formed the opinion that the white ants have rendered valuable service to Egypt by amassing the fertile mud which is carried down to the Nile Delta, and venture to put forward the following considerations in support of the theory. On entering the Abyssinian borderland, one cannot help remarking the structure of the ground. Our track lay generally along watercourses, and the beds of dry mountain torrents. Here the soil was composed of sand, shingle, and pebbles. On either bank, and for some distance beyond, lay an expanse of basaltic stone, on which little earth was to be seen. Grass grew there, but by no means so abundantly as in the plains, and there were climbing plants, such as convolvulus, and ivy, which clung to nearly every tree. These creepers made the path extremely difficult to follow. Further away from the water-courses lesser vegetation and longer grass appeared. The white ants' nests, from nine to twelve feet in height, were found here, usually close to a soft wooded tree. The roots of it, in most cases, had been attacked by the insects, and converted into "white ant earth." The trunk afterwards undergoes the same process, and by the advent of the rainy season only the outlying twigs remain intact. A heavy gust of wind will then overthrow the *simulacrum* of a tree. The rain falls in torrents, and the compost which the insects have made of the timber is broken up and carried by innumerable channels into the tributaries of the Atbara, and finally reaches the main stream. It is well known that the Atbara brings down the greatest quantity of this mud, the Blue Nile carrying less, and the White Nile, the most sluggish stream of the three, least. The two more rapid rivers rise in Abyssinia, in regions where the white ant is extremely destructive to vegetation. Moreover, the innumerable ant-heaps * are made entirely of

* That most resourceful of travellers, Sir Samuel Baker, once put an ant-hill of this kind to a singular use. The incident may be new to some of my readers. The explorer wished to make soap in the Soudan. "I had neither lime nor potash, but I shortly procured both. The hegleek-tree (*Balanites Egyptiaca*) was extremely rich in potash; therefore I burned a large quantity, and made a strong ley with the ashes; this I concentrated by boiling. There was no lime-stone; but the river produced a plentiful supply of large oyster-shells, that, if burned, would

earth, which crumbles under the rains, and is swept into the watercourses in the season of floods. And in the dry months the ants—besides devouring straw and the bark of living trees *—eat into every branch and twig that the past storms have brought to the ground. Carrying earth into their excavations, they hollow the wood, as in the case of the soft standing timber which they attack, and leave just the outer covering intact. I have often stooped to pick up a stick for the cook's fire, and found that my fingers broke a thin shell of bark, and scattered the contents. Another circumstance which seemed worthy of remark was that we found no deposit of "Nile mud" within the Abyssinian boundary, nor any earth resembling it; so one may conclude that, as soon as the Atbara, when in flood, reaches the level of the Soudan, it spreads a deposit of this fertile soil beside its banks, though the great mass is carried into the Nile. According to this view, the fertility of the alluvial districts of Lower Egypt is in no slight measure due to *detritus* from Abyssinia, and the white ants have contributed an important share to the resources of the Nile valley. Of course, I do not for an instant contend that all the mud deposited by the river in Egypt is

yield excellent lime. Accordingly I constructed a kiln, with the assistance of the white ants. The country was infested with these creatures, which had erected their dwellings in all directions; these were formed of clay so thoroughly cemented by a glutinous preparation of the insects, that it was harder than sun-baked brick. I selected an egg-shaped hill, and cut off the top, exactly as we take off the slice from an egg. My Tokrooris then worked hard, and with a hoe and their lances, they hollowed it out to the base, in spite of the attacks of the ants, which punished the legs of the intruders considerably. I now made a draught-hole from the outside base, at right-angles with the bottom of the hollow cone. My kiln was perfect." ("The Nile Tributaries of Abyssinia," pp. 424, 425.)

* I have seen living trees of the height of ninety feet, the bark of which had been entirely converted into a layer of "white ant earth."

supplied by the white ants, but I believe that the wonderful productive property of the alluvial deposits is due to the work of these insects in the western borderland of Abyssinia. Therefore, in my belief, the white ant has justified its existence, though its room is undeniably preferable to its company.

On January 3 I saw a buffalo, tried to stalk it, and lost it. While I was on the way to rejoin our party, a water-buck sprang up within range—and so startled me, that he got away before I took an aim.

That day we marched seven hours, and covered twenty miles. Our camping-place was in the rocky course which a torrent follows in the rainy season. There were pools in it, and we had an abundance of clear water.

The aneroid barometer showed that we were now three thousand feet above the sea, and the air was chill as soon as the sun went down. It was the first day on which we had reached any considerable height. At this altitude bamboos form the prominent and characteristic vegetation.

On January 4 we continued our ascent. The narrow path followed a zig-zag course up a steep mountain-side. The track was full of loose stones, and we constantly had to "negotiate" boulders and big rocks or scramble through cramped passes. In these the larger loads stuck, and it will be readily inferred that we made slow progress. I heard of no other practicable route through this region, and believe that it would be well-nigh impossible to carry heavy baggage this way.

We reached the summit at last, and saw in front of us, in the distance, the plateau in which Lake Tsana lies. We were almost on a level with it. Far away, many mountains rose in view, clearly outlined, and showing light and shade in a soft and lovely purplish blue colour. There was no glacial cap upon these high peaks, though it is probable that the loftiest points in Abyssinia are within the line of perpetual snow.*

One of our escort of Habashes from the Arab Battalion pointed eagerly to the high lands opposite, and said to a Soudanese boy who was standing near, "Look at my beautiful country!" The soldier's manner showed all the zest of a schoolboy returning home for the holidays. The answer was, "Call this a country! Where are the people?" Our men from Upper Egypt were puzzled and disdainful. They had travelled eighty-six miles in Abyssinia and had not seen a dwelling or a sign of cultivation.

A deep ravine lay below us, and we descended through it and entered a great gorge which opened out as we advanced. We were on rocky ground, covered with a layer of earth and loose stones. Our track wound through dense groves of bamboos, and we had to perform gymnastics on muleback to avoid a whipping from the canes. In this uncomfortable covert we passed a party of traders taking coffee to Gallabat. They had with them about sixty loads.

Presently our convoy, which in extended order covered nearly a mile and a half, debouched upon the

* "My sight was now delighted by frequent views of the summits of the Simyen mountains, thickly covered with snow. As this was the month of April, one of the hottest, I can easily believe that some of these mountains are never free, but are within the region of perpetual winter. They are the highest in Abyssinia, attaining an elevation of fifteen thousand feet in Aboo Yared, the loftiest peak." (Dufton, "A Journey through Abyssinia," p. 196.) The heights of Simyen lie north-east of Lake Tsana, between the lake and Adowa.

valley of the river Geerar.* The bed, which is of bare rock, was dry when we passed through it. I saw here many basins neatly excavated in a curious fashion. They varied in size, but at the bottom of each was a stone. Evidently, this had first been lodged during the time of flood in some small hollow in the water-course. There, being constantly driven round by the eddying water, it had worn a deeper and deeper hole, at the bottom of which it rested in the dry season. We followed the course of the river up stream for about five miles.

The valleys on either side are sharply cut ravines in the mountain-sides. They were covered with grass at that period of the year.

We lunched in the shade of the river bank, and afterwards visited some hot springs which were near at hand. The first was a runnel of clear water guided from its source in the rock to a pool by a wooden gutter. I put my hand under this, and judged that the temperature was well over 150° F. At a little distance we found another spring which, welling from the ground, filled a basin about forty feet in circumference and one foot deep. Here, too, the temperature of the water was high, but not so high as to prevent bathing, and a roof of thatch had been built over the pool. We saw several Habashes, men, women, and children, who inhabited some rough, small thatched huts which had been put up close by. Some were old dwellings, some newly constructed. I could not ascertain whether these people resorted to the springs for a "cure"

* "The Geerar is a typical mountain river in a country consisting almost wholly of impervious basaltic and granite rocks; it is evidently a furious torrent after rain, but dry except for the merest trickle at other times; it runs in a deep narrow valley, shut in, hot, and airless, and probably very unhealthy." (Sir W. Garstin, "Report upon the Basin of the Upper Nile," 1904.)

or merely formed a settlement established by keepers in charge of the place. Apparently all belonged to one family. I visited their camping-ground, but when I approached the women and children bolted into one of the huts. Perhaps they had not seen a European before and did not consider the first example prepossessing; or they may still have supposed—as most Abyssinians at one time did—that all persons who came from that continent were Turks. In the latter case I excuse their timidity. They watched me through the thatched walls of the hovel just as a rat sometimes watches a suspect from the entrance of its hole.

The water, as far as I could judge, was pure. It was tasteless, and very "soft," as I found by using it in my bath. No steam was visible at the springs, though before I left them—about four in the afternoon—the air was becoming cool. Some large trees grew near the pool, and I saw a big and fine monkey with a shaggy face and a growth of fluffy white hair on its tail leaping and squatting among the boughs.

There are many hot springs in Abyssinia, and they are highly valued by the natives for the medicinal qualities attributed to them. Probably in some places the virtue of the waters is real, in others merely reputed.

The most frequented baths are at Wansage on the River Gumara, which flows into Lake Tsana. Dr. Stecker visited the spot in 1881, and wrote the following brief account of it,* which throws a light on Abyssinian social customs: "The hot spring issues from the left bank of the river and rises to a height of two or three metres from the ground. It fills a basin constructed by King Theodore with water

* "Mittheilungen der Afrikanischen Gesellschaft in Deutschland," vol. iii., 1881.

of a temperature of 37° C. A hut has been built over the basin"—as at the spring which I saw—"and the Abyssinians who are trying the 'cure' here sport in the water all day long. In arrangements these baths resemble Ostend and Trouville in miniature, seeing that women and men, youths and girls, spend their time in the bath in lively promiscuity and keep up an intercourse that is not always decorous. Quarrels often arise among the 'cure' patients, especially when some one has used the baths longer than is permitted to him. Thus from early till late one hears the loud tones of the brawlers and the lamentations of women and children, who often come in for a beating on these occasions."

On January 5 our road lay at first beside the River Geerar, which is the main source of the Rahad. Presently it diverged from the bank, and the track, which lay over stiff hills, became worse. At some points the donkeys could not pass, and we had to lift them bodily, loads and all, over obstructions, or push them past the rocks between which their loads were jammed. Once the animal that was carrying Dupuis' valise and Gladstone bag, half slipped into a steep ravine with water at the bottom, forty feet below. In another place I saw an unlucky little beast turn two complete somersaults while it tried to scramble down a sloping ledge of rock. To my horror I caught sight of my medical bag on the top of its burden. I rushed up, and helped to unload the donkey, which seemed none the worse for the fall, and, to my astonishment and relief, I found all my doctor's gear intact. Even the tabloid bottles, with a precious stock of quinine and phenacetine in them, had escaped by a miracle.

In crossing and recrossing the river-bed I was able to note the height to which the floods rise in the water-courses. I saw that flotsam, such as wisps of straw and bits of wood, had been lodged in branches and twigs at least twelve feet above the bed of the stream, and this, in the upper reaches among the hills, has a width of twenty-five to thirty feet on an average.

We camped at the foot of the mountain-side which would give us our final climb to the lake-plateau. Johannes pointed out the route to me, and I thought that one would have as much chance of shoving a donkey up the Great Pyramid as up that sheer precipice.

Some Habashes came to us at the camping-ground, bringing a cow, a chicken and ten eggs. They told us, through Johannes, that the cow was "a present," and he said that it was worth ten dollars.* We decided to

* These are the "Maria Theresa" dollars. "The Maria Theresa dollar is a facsimile of the coin of 1780. It is still issued by the Austrian Government, and circulates as a trade coin nearly all over Africa. Though roughly about the size of a five-shilling piece, it only exchanges for a little more than its intrinsic value, which was about 1*s.* 11*d.* when I left Africa." ("Abyssinia," by Herbert Vivian, 1901, p. 239.) Dufton wrote, with regard to Abyssinian marketing: "The Maria Theresa dollar is the only coin used in these transactions under the name of *bir* (silver). This piece must be of a certain kind, however, containing the spots which form the queen's tiara, and other marks clearly defined; for which reason it is called by the Arabs of the Soudan Abou Nukter, or father of spots." ("A Journey through Abyssinia," p. 61.) Mr. Vivian gives the following table of Abyssinian currency:—

4 cartridges	=	1 salt
4 salts	=	1 Maria Theresa dollar

The "salt" is a bar of that mineral, which is ordinarily carried and used in lieu of coin. I found that its value was, roughly, sixpence, as Mr. Vivian says. At an earlier date its purchasing power seems to have been less. "Pieces of salt, called *tsho,* are the monetary medium throughout the greater part of Abyssinia. This salt is obtained from extensive mines in the country of the Danakil, where it is cut into small blocks, eight inches long by one inch and a half in breadth, each of which is of a variable market value of between twopence and threepence." ("A Journey through Abyssinia," pp. 61, 62.) Evidently there has been "appreciation of salt" in the land of the Negus since 1862, when Mr. Dufton visited it.

RAIN COUNTRY. [See p. 64.

offer a "present" of fifteen dollars in return, and counted out the money. To our astonishment, the Habashes marched the cow off, saying that they would not accept the gift. We picked up our coins, feeling rather small. But money is "a good soldier and will on," as Falstaff had found. In an hour the animal was brought back, and the "presents" were duly exchanged. Then there was a wrangle among our Mohammedans as to the ceremonial proper to be observed in slaying the unlucky cow; for some of the boys were of one sect, some of another. The question was settled after much haggling, and they held high festival and ate meat late into the night. A few choice cuts had been reserved for ourselves, but the bulk of the carcase was handed over to our followers, and they left only niggardly fare for the vultures next morning.

A very good impression of the character of the country through which we had been marching is given by Mr. Dufton's short description of it. "The road was very uneven, now ascending a steep mountain-side, now descending into a deep valley. The country was magnificent, far surpassing anything I had previously seen. The high mountains of the Scotch highlands, covered with the fertility of the Rhineland, would best represent it, but the vegetation was of a nature quite different from that of the Rhine, characterized as it was by the luxuriance of the tropics. Once the road skirted the side of a mountain the summit of which, raised one thousand feet above our heads, looked down into a deep valley another thousand feet below our feet. On the opposite side of the valley the land rose to a similarly steep eminence, which, in one part, was connected with that on which we stood by a low chain of undulating ground, so that a pretty little stream at the bottom, like a silver thread in the dark shadow of the mountains, wound about searching for its channel. Fruitful fields hung over it thick at every curve. The hills, of secondary formation, were broken here and there into rocky chasms, through which leaped innumerable falls of water in their downward course to join the stream; and here I saw for the first time the beautiful Euphorbia called the Kolquol, whose dark, candelabra-shaped branches, tipped with bright orange-red flowers, stood out in deep relief from the lighter ground. Bright flowers of every variety, most of which were unknown to me, but amongst others the familiar wild rose, the honeysuckle and the jessamine, lent their beauty and fragrance to the scene." *

The knotted sapling which I photographed in the course of this day's journey (January 5) has an historical interest which tempts me to moralize on the variability of human fortunes and the happy tranquillity of vegetable life. The tree had been growing in this distorted form since the time of the battle of Gallabat, which was fought between the Dervishes and the Abyssinians in 1889. I have alluded elsewhere to the singular vicissitudes of the fight.† The Dervishes, in light order, pursued their foes as far as the lake district, a circumstance which shows that raids are possible even in this difficult country. The Mahdi's men, like our own War Office on a more recent occasion, had extremely little geographical or topographical information. So the advance party twisted the saplings to mark the route for those who came after them, and to guide the force on its return journey to the Soudan. The trees are the only remaining memorial of the Dervish raid.

* "A Journey through Abyssinia," pp. 47, 48.

† See p. 179.

THE KNOTTED SAPLING NOW BECOME A TREE. [*See p.* 66.

In this region I saw very few birds. But on the night of the 4th, while I was getting ready for bed, I heard one whose notes ascended through a perfect chromatic scale. My friend Dupuis told me that these songsters are common in India, where they are called "brain-fever birds." Whether he spoke as a humourist or a genuine informant, I cannot say. In the Soudan I had often heard a bird whose notes reversed the process and *descended* the chromatic scale very perfectly. I omitted to suggest to my companion that he should import some of this species to India and try its performance as a remedy for brain fever.

On the morning of January 6, we started to climb to the plateau in which Lake Tsana lies. The ascent commenced immediately. The narrow track was extremely steep, and, as on the previous day, our path was full of loose stones and led us over great rocks that crop out of the mountain-side. The donkeys were constantly slipping and falling. Some came to a standstill, and refused to budge. We had to shove the animals by main force over boulders and up slippery ledges of rock, and at places not a few of them were raised bodily, loads and all, by means of their tails and forelegs, and lifted over obstacles. At this rate of ascending we covered two miles in three hours and a half, and still had a climb of another couple of miles ahead of us. Then the character of the track changed, and we travelled round a horseshoe-shaped chasm, following a path four feet wide, with a sheer precipice four hundred feet deep below us and another rising to the same height above our heads. It was a fine sight, and there is a lovely growth of cactus on the mountain side. Besides, we looked out upon a vast expanse of beautiful scenery, but I felt uncomfortably like the ungodly of whom the psalmist said that they were set in slippery places with a great risk of being cast down and destroyed. "Oh, how suddenly do they consume; perish and come to a fearful end!" * It was an inopportune moment to recall the text.

We reached the plateau at last, after another stiff climb upward from the chasm. I need hardly say that the donkeys were utterly fagged out. We had left the mimosas and the bamboos below us, and Dupuis' aneroid barometer showed that we were some six thousand feet above sea-level. Here many species of cactus, large and small, abounded. The soil on the plateau is rich, and the ground was thickly covered with lush plants in blossom. I noticed, as I passed, the familiar "red-hot poker," the wild strawberry, moss of many hues growing luxuriantly, the maiden-hair fern, and, on the trunk of a dead tree, the Tonbridge fern. Many springs were bubbling from the rock, and their courses were marked by the tenderest and brightest tints of this wild mountain garden. I found that Dupuis, who had marched at the head of the column, had been stopped by a couple of soldiers on the edge of the plateau. The Abyssinian "regular" has no uniform, but wears a dirty shama and the rest of the national costume, and carries a rifle of an obsolete French pattern. These men were not acting under Menelek's orders but had been sent by the deputy of the chief of the village, who was himself absent upon a visit to Ras Gouksha, one of the great feudatories of Western Abyssinia. They said that they had orders to stop the three Englishmen, and I found that Dupuis had arranged to lunch at that spot and await developments.

Presently the "Deputy-Governor" arrived. He was barefoot, and his shama and linen trousers showed that

* Psalm lxxiii. 18.

MESSENGER SENT TO STOP US GOING DOWN TO THE LAKE. [*See p.* 68.

In Abyssinian intercourse the raising of the shama to cover the nose is an assertion of *superiority* by the person who so acts; to raise the shama to cover the mouth implies a claim to *equality*. To lower it to the waist is a sign of extreme deference; strictly, it implies servitude.

he was a sound observer of Abyssinian custom in respect of cleanliness. He was escorted by some grimy fellows with forbidding faces who carried guns. The "Deputy" bowed low and shook hands with the three Europeans. Then Dupuis offered him a camp-stool for a seat and talked to him, Johannes interpreting. Menelek's letter, giving us the right to pass freely through any part of his realm and calling on his lieges to assist us, was read, but to our consternation the "Deputy" refused to let us go down to the lake, which was eight miles distant, until we had received permission from his Ras.* Three days, he said, would elapse before this could be obtained. We gave him a drink of green chartreuse, and then he promised that he would allot a camping-ground to us about a quarter of a mile from Lake Tsana, but insisted that we must not pitch our camp on the shore.

After the palaver he remained to watch us. A skin had been spread for him under a tree a few yards distant from us, and he lounged on this, staring at us when we were near, and following all our movements as closely as he could without giving himself trouble. Long grass grew around our camp, and I noticed some tamarind trees. We strolled among these surroundings with our guns and shot a few wild pigeons for dinner. Great numbers of these birds are seen on the high ground. Then, *more anglicano*, we had afternoon tea.

After sunset the air was very keen, and we put on our thickest clothes. The variation between the heat of the day and the cold at night is extreme on the plateau, and of course it is sharply felt. We turned in early and slept in the open. Our guard lay around us; for we

* Chieftain. As in the feudal times in Europe there is a chain of allegiance—often enough broken—between the king and the mass of his subjects. This Ras was the "Deputy's" feudal chief.

were in strange company. Besides the "Deputy," who might be calculating whether it would pay him better to see what present we should offer and avoid the risk of complications, or to cut our throats as quietly as possible and loot the camp, we had as near neighbours a party of Abyssinian traders on their way to Gallabat with coffee. They had bivouacked under a tree close by. They might or might not be disposed to share in the pillaging, if any took place. In the land of the Negus it is well to remember that "no one is expected to feel ashamed of any crime or vice; and whereas in other countries men in committing serious crimes are morbidly excited, in Abyssinia they are perpetrated with indifference, and generally recounted, sometimes by the individual himself, certainly by others, with gaiety and laughter. . . . Theft is in many provinces regarded as an honourable employment; highway robbery is quite excusable, even if accompanied by homicide." *

After all, we slept as peacefully as if we had been in Anerley or Tooting. On the morning of January 8, we rose when the sun had warmed the air, and ignored the presence of our dirty warder, who still kept us under observation. While we were breakfasting I saw a white umbrella approaching through the long grass, and shortly the priest of the village hard by came in view. Neither the hamlet nor its church was visible from our camp. This divine did not wear a turban, which is the emblem of priestly rank in Abyssinia, but had on his head the old straw lining of a tarboush. His other visible garment consisted of a length of yellow and plum-coloured chintz. He carried a staff surmounted

* Consul Plowden (1848–60), quoted in "Abyssinnia and its people," by J. C. Hotten, 1868, pp. 130, 135.

by a cross of filigreed iron. One attendant held over him the white umbrella, which had a blue lining, another bore before him an open book, on one page of which appeared a picture—the quaintly stiff and gaudy depiction of saints and sacred persons which is an unvarying convention of Abyssinian art—and on the other page I saw manuscript in the Amharic or Geez character.

The "Deputy" approached the priest, bowed, and kissed the book, and some of his retainers followed his example. Then Johannes explained to us that it was customary for travellers to make an offering to the church of this village, which is on the boundary of inhabited Abyssinia. Dupuis asked what the usual oblation was, and the "spiritual pastor" had the effrontery to reply, through Johannes, that it was ten dollars. Upon this Dupuis remarked that we had been very badly received and were dissatisfied with our welcome, that hindrances had been put in our way in spite of the terms of the king's letter, and that he should give no alms to any one.

This declaration disconcerted both the priest and the "Deputy." The former departed without further parley. The latter continued to follow us and spy upon us. Dupuis and Crawley decided to disregard his prohibition and advance. They moved in front of the column, and took with them Johannes, his two attendants, and eight soldiers. I brought up the rear with a couple of mounted men for escort. My progress was not interrupted, and gradually Lake Tsana came clearly into view. I found, when I rejoined my companions at the camping-ground which they had chosen on the shore, that the "Deputy" had made an attempt to stop them when they were about five miles from the spot. He had given them to understand that they might see the lake, but must not go down to the water's edge. Dupuis replied that he should regard no such order, and marched ahead. Upon this the "Deputy" shrugged his shoulders, and forthwith took his leave, saying that he must then return to his house, but that he would visit us in the evening, and supposed he would find us on the shore of the lake, as we were resolved to proceed thither. He had an answer in the affirmative, and departed, to the relief of the whole party.

CHAPTER VI

Lake Tsana lay before me as a vast expanse of blue water stretching to the horizon. The shore here, near the village of Delgi at the north of the lake,* is in places sandy, in others covered with grass. At some points one can walk to the edge of the clear water, at others the shelving bank leads, by a scarcely perceptible incline, to stretches of reed-grown, swampy ground. I saw no water-weed upon the surface.

The lake lies at an altitude of 1942 metres (6372 feet) above the sea-level. Its length from the mouth of the Magetsch to the outlet of the Blue Nile is approximately forty-five miles, and its breadth on the twelfth parallel of longitude is about thirty-seven miles. The map published in this volume is mainly reproduced from that prepared by Dr. Stecker for the African Association of Germany. How painstaking he was in collecting information may be inferred from the fact that during his excursions in the native boats he took three hundred soundings—an example of patient research which those will best appreciate who have seen a "boat" of the kind. Dr. Stecker gave the following results of his survey:—"Lake Tsana covers a superficial area of 2980 square kilometres. All the islands together possess a superficial area of about fifty square kilometres, of which forty are included in Dek, and four

* See map.

in Dega. I found the greatest depth between the island of Dega and Zegi, viz. seventy-two metres, while the deepest sounding between Korata and Zegi was sixty-seven metres. Between Dek and Adina the depth ranged between thirty-two and forty-seven metres. But I am fully convinced that the deepest places—in my opinion having a much greater depth than one hundred metres—are to be found north of Dek in the direction of Dega and Gorgora. One cannot, however, well venture to make an excursion to those parts in the fragile Abyssinian craft." *

This beautiful lake is everywhere girt by mountains, and in some places they rise directly from its shore. As a consequence it is exposed to sudden gusts and squalls that sweep down the valleys and ravines, and these would make navigation risky for small sailing boats. Hippopotami swarm in the waters. The Abyssinians hereabouts—Christians and Mussulmans alike—regard their flesh as unclean; but the hippos do not "lead a very snug life," as Dr. Stecker supposed. They are constantly harassed by the natives, who shoot them for the sake of their hides. From these are made the whips called *coorbatch*, which are in general demand and are much valued. Plowden wrote that "the crocodile, that is found in most of the large rivers of Abyssinia, does not infest this lake." † So far as my personal observation goes, he was right. But, knowing the enterprising character of the Blue Nile crocodile, I should think it wise to verify the statement, while on the spot, by careful observation.‡

* "Mittheilungen der Afrikanischen Gesellschaft in Deutschland," vol. iii. p. 32.

† "Abyssinia and its People," by J. C. Hotten, 1868, p. 128.

‡ Mr. MacMillan, describing some of the incidents of his recent

I did not hear of the " Deputy's " peaceful departure immediately on arriving at the lakeside, for neither of my comrades was in view. After a few minutes I saw one strolling towards me carrying a wild goose, which he had just shot with a rifle, and then the other appeared, coming from the opposite direction, with a brace of wild duck. It was clear that the lake country was well stocked, and that the expedition would fare sumptuously if it were not cut up.

All our tents were pitched by eleven o'clock, and we began to think of lunch. A man looks forward cheerfully to his meals in the fine mountain air. But on that morning anxiety about the future harassed our minds. However, I had no leisure to make forecasts, for I had to attend to one of our soldiers who had fever, and one of our boys who had dysentery, and then I found that I was appointed honorary surgeon to the neighbourhood, and that a patient was waiting. He was an old man, who had a large abscess in the sole of his foot, and had endured the affliction for more than a week. I operated, while his friends looked on. There was no superfluity of medical stores, and it seemed fair to make " the case " supply his own bandage. This, when

expedition to the upper reaches of the Blue Nile, told an interviewer: " We were roused at four in the morning by the wild screams of one of our people (a Somali), who had been seized by a big crocodile, which had crawled up the rocks. The brute had seized his victim by the head and dragged him to the water's edge. By throwing stones at the crocodile we succeeded in making him leave the Somali, who, though terribly mauled, eventually recovered."—*Egyptian Gazette*, October 21, 1903. Mr. MacMillan told me subsequently in conversation that the Somali, when he was attacked, was sleeping with his head on his arm, and that the crocodile seized both head and arm, with the result that the Somali put his finger in the crocodile's eye. Mr. MacMillan was inclined to believe that this action made the reptile release the man. The guns and ammunition of the expedition had all been lost, and were at the bottom of the river, and the night was pitch dark.

produced, proved to be a piece of dirty shirting. It served, over sublimate wool, and the patient was very grateful and thanked me profusely.

Parkyns made some striking and interesting observations about the insensibility to pain displayed by the Abyssinians and the African races of all kinds.* He said, "I have never noticed in Africa any education for the purpose of rendering men patient under suffering," and he attributed the power of endurance to the hardening effect of a rough, primitive life. How far this is the cause and how far the advantage may be due to a somewhat lower form of nervous organization than that developed in Europeans I will not attempt to determine. But surprising as are the facts related by Parkyns, I do not doubt that he has described them without exaggeration, and my own experience showed that the Habashes stood pain well, though I observed no inclination towards heroism among them.

When I had finished the public demonstration in surgery, I noticed a small throng of people around Dupuis's tent. Their faces displayed their satisfaction, and I joined the group to learn what had happened. Then I heard that the true representative of Ras Gouksha had arrived, and that the "Deputy" was a fraud. The envoy who had now come into our camp had expected us to reach the lake by another track, and had taken his post to await us. This man greeted us with a most civil welcome, gave us teff,† eggs, milk, fowls—in fact,

* "Life in Abyssinia," pp. 383 *seq.*

† Native bread. "The chief corn of the country is 'teff' and 'dágousha,' if, indeed, we may venture to include these under the head of corn; for they both resemble different sorts of grass, and the seed is not larger than rape or canary seed. Of each of these there are various qualities, esteemed according to their colour, white, red, or black. White

all the supplies which we chiefly needed—and offered to accompany us round the lake and see that we were everywhere treated with courtesy. He seemed a smart and "likely" fellow, and any one may imagine our delight in the prompt exposure of the "Deputy" and the dramatic change for the better in our situation. We gave our new friend a stiff drink of green chartreuse, which he swallowed at one gulp. Then he went off to collect further supplies for us.

A little later the "Deputy" returned to exculpate himself. He grovelled on the ground, imploring forgiveness. He had brought a sheep as a peace-offering, and when we had added this to our belongings, we graciously pardoned him. Upon hearing that he was absolved from his guilt, he stooped down and kissed a stone close to Dupuis's feet.

We had had enough parley for the day, so Dupuis and I started upon an explorers' tramp, carrying our guns. We walked about two miles in a north-westerly direction, and saw innumerable kinds of water-fowl along the margin of the lake. The birds were scarcely shy of us at all, and we approached within ten yards of wild geese before they rose. I noticed among the mass a species of goose like a big Muscovy duck, with dark green plumage and white feathers in the wings, plovers of all sorts, herons, pelicans, snake-birds, and the ibis in numbers. We shot none, for we had meat enough in store, and it would have been sheer slaughter to do so. I did, indeed, try my luck with some quail

'teff' bread is preferred by all natives even to wheat bread. . . . The 'teff' is considered by the Abyssinians wholesome and digestible; but so far from being satisfied of this, I am doubtful of its containing much nutritious property; and as for its taste, only fancy yourself chewing a piece of sour sponge, and you will have a good idea of what is considered the best bread in Abyssinia." (Parkyns, "Life in Abyssinia," pp. 198–200.)

that took to the wing, but no harm came to them, and I was not sorry.

We returned to the village. It was pleasant to see kine again after the long march through the deserted border-country. They are here of the long-horned, hump-backed "Zebu" kind common throughout Africa and the East.* My friend Crawley had taken a rod and line to the edge of the lake. He landed a fish about 2 lb. weight, of the perch tribe. So we had a very complete *menu.* The cold was keen after sundown, but we had a cheery finish to our first day by the lake, after all. And it would hardly be possible to look upon a more lovely scene than that which we saw from our camp when the moon had risen.

We were not astir very early in the morning of January 8, and dawdled over breakfast until half-past seven; it is almost worth while to live a strenuous life at times in order to enjoy dawdling afterwards. Dupuis and Crawley set out with an assortment of gear to make a survey, and I amused myself with a butterfly-net and a camera.

Then I inspected my patients. Several came into camp from the neighbourhood, as I had expected. The old man with the abscess presented himself, and I found that the injured foot was healing well. He thanked me again very profusely, and I believe that after this second visit he departed to the village whence he came with a good opinion of European surgery. Another Habash found me less satisfactory as a physician. He had come to ask what medicine he could

* In other parts of Abyssinia "there are also two kinds, one a very large one and another of Jersey size, that show absolutely no trace of an Eastern origin, and are as straight-backed as any English short-horn." ("Modern Abyssinia," by Augustus Wylde. Methuen, 1901.)

HOUSES AT DELGI. [*See p.* 79.

WASHING OUT 'TEDJ' POTS AT DELGI. [*See p.* 79.

take to cure the headache caused by *tedj*. Now, *tedj* is the beer, or mead, of the country; it is made from fermented barley, and flavoured with honey diluted in the proportion of one part to three parts of water. It is a very heady—and, to Europeans, a most nasty—drink, and the Abyssinians consume enormous quantities of it. Parkyns was told of a man who was said to have swallowed twenty-six pints at a sitting, on the occasion of a wedding-feast at which the English traveller was present. But he regarded this statement as "a stretcher." * I told the inquirer that the one and only prescription was not to drink *tedj*, and thereupon the little audience of his fellow-countrymen enjoyed a laugh at his expense.†

I was glad to see that the servants set to work to wash their clothes with soap in the lake. The cleansing was needed. Soon the tents in our camp were draped with garments enough to occupy the wash-lines of a

* "Life in Abyssinia," p. 259.

† I do not know whether the women drink much *tedj*, but even the ladies of the land did so in Bruce's time. He was in Gondar when "Ozoro Esther's sister, the Iteghe's" (queen-mother's) "youngest daughter was married to Powussen, the governor of Begemder. The king gave her large districts of land in that province, and Ras Michael a large portion of gold, muskets, cattle, and horses." On this occasion "the Ras, Ozoro Esther, and Ozoro Altash entertained all Gondar. A vast number of cattle were slaughtered every day, and the whole town was one great market: the common people, in every street, appeared laden with pieces of raw beef, while drink circulated in like profusion. The Ras insisted upon Bruce's dining with him every day. After dinner they slipped away to parties of ladies, where anarchy prevailed as completely as at the house of the Ras. All the married women ate, drank, and smoked like the men; in fact it is impossible to convey to the English reader, in terms of proper decency, any idea of this bacchanalian scene" (Head's "Life of Bruce," pp. 297, 298). In Parkyns's time a wedding-feast was an orgy of the same kind for all classes and both sexes. And as there is a complete absence of *gêne* in the conduct of modern Abyssinian women in other respects, I have little doubt that they still favour the *tedj* when the mood prompts them.

whole suburb. The lake had a pleasant temperature for bathing, and the men stayed in the water till their clothes were dry.

On this morning we had a visit of ceremony from another priest, whose umbrella was of many colours. His attendants were a boy dressed in plum-coloured chintz with a yellow scroll-pattern on it, another boy, who was naked, and carried a bell which he tinkled incessantly, and three Abyssinian students. These learn to read and write the Gheez language, and I think the Bible is the only book which they study. They are lads from the villages who are "candidates for orders," and the theological classes are held in the church-porches. I secured two satisfactory snap-shots of the priest. He received an oblation of five dollars with an absolutely impassive face, and then left us in doubt whether he was secretly gratified by the amount of the offering or inwardly disgusted by it.

Dupuis and Crawley resumed their survey work in the afternoon, and I strolled away from camp with my gun and brought down a lesser bustard, of the size of a turkey. I also shot a brace of quail, but lost them in the long grass. Altogether it was a quiet day. The climate was now very pleasant, neither too hot at noon nor too cold at night. In the early morning there was a dead calm. After this a little breeze came down from the north-west, and the wind remained in that quarter till the afternoon. At four o'clock it shifted right round to the south-east, and blew pretty stiffly about seven. We always saw lightning in the evening. It seemed to play over the lake.

On the morning of the 9th Dupuis and I got our Berthon collapsible boat ready. This had caused us much trouble on the upward march, as it is awkwardly shaped for donkey-transport. We took our guns and angling tackle, and paddled about three miles westward. Then we landed on a sandy beach, left two of our boys in charge of the boat, and went in search of game, taking one man with us. We had not walked far when we sighted a covey of guinea-fowl. We got ahead of them, and were trying to drive them towards the water when about forty rose in all directions, and we could not re-load fast enough. Several were lost in the thick grass, chiefly owing to the boy's stupidity; he could not make up his mind which bird to pounce upon first. As we were returning to the beach by the way we had come, Dupuis put up a brace of partridges and shot them. They must have "sat tight" all through the fusillade, and only rose when my companion nearly walked them down. We carried about ten head back to the boat.

When we came to the shore we found our two men in a state of great excitement. As soon as the guns were fired a hippopotamus that had been asleep about ten yards from them sprang up, and rushed into the water, snorting. The beast had nearly frightened them out of their wits. Looking at the path I had followed after landing I saw that I must have passed within a yard of the place where the hippo lay. The men now drew my attention to something in a tree, but I could not discern at first what they were pointing out. After peering for three or four minutes I saw a pair of gleaming eyes above a branch in the deep shade. I fired with number four shot—the only size I had—at a distance of about ten yards, and down came a civet cat. It was hit in the head and dead, when I picked it up.

We tried our luck with the fish while the men paddled us back to camp, but caught nothing. I set about preserving the skin of the cat after lunch, rubbed some arsenical soap well in, and packed the hide in an empty cigarette tin. It was brought to England in good condition. While I was busy in this way, my comrades were taking soundings of the lake from the boat. Later we went out together with our guns, and added two bustards, three partridges, and a quail to the stock in the larder.

During the day, whenever I was in camp, I was importuned by patients; many had trivial ailments, and others troubles, such as chronic ophthalmia, which I certainly could not cure during a three days' sojourn at Delgi. If they judged that I was not sufficiently moved by the account which they gave at first of their malady, they described what they suffered from another, and a fictitious one. Thus, if I told a man who had chronic ophthalmia that I could not help him, he would remain to declare that he was consumptive, or that his feet needed treatment. They became so tiresome at last that I was thankful for the prospect of resuming the march next day. It seems to be taken for granted in all countries that a doctor ought to be more long-suffering than any other person.

On the morning of the 10th I received, by Abyssinian post, a letter which had been despatched from Port Said on November 25, the day on which I left Cairo. It reached me by a roundabout road; for it had been to Wady Halfa, Berber, Kassala, Sennaar, and Gallabat. I was surprised that it came to hand. The postman was a Soudanese black, and, when we met him, he was on the return journey to Gallabat bearing some official communication. These couriers carry before them—like a wand of office—a long cane, which is split at the top. The "mail" is inserted in the orifice. The cane is a badge that is respected, and I was told that the letter-carriers are never stopped. They amble along at a steady jog-trot, and cover a great deal of ground in a day.

We made a short journey of about ten miles to the eastward after leaving Delgi. Our road kept us some little distance from the northern shore of the lake, and took us through long jungle grass, so tall that we could not see over it, and so thick that there was no possibility of using a gun in it. We camped on a spot where this grass had been beaten down, and spent the afternoon reading. There were no mosquitoes, but the flies swarmed into our tents, and we sympathized with Pharaoh and the Egyptians.

Several of the boys from the Soudanese lowlands had a touch of fever on this high ground. Three were on my hands that day. Quinine was a quick and sure remedy for them. I think that they were affected by the great change of temperature between midday and nightfall much more than Europeans.

The morning of Sunday, January 11, was very cold, and the thermometer stood at 31° F. just before sunrise. I noticed that it always rose to a point above 80° about noon, and these figures will give an idea of the effect of the variations. The servants were shivering when they served our breakfast, but at half-past seven we took off our overcoats, and when we started at a quarter-past eight the day was warm. Our road lay through tall grass, and took us over undulating country, and then up a long, gradually rising slope to Tschenkar, where we arrived at one o'clock. This was by far the largest village that we had yet seen in Abyssinia.

Tschenkar is four miles distant from the lake. As we travelled we had skirted the base of the promontory called Gorgora, which lay on our right hand. This is very lofty ground. Stecker ascended to the highest point of it, the summit of the mountain called Goraf. He found by barometrical measurement that this was 2134 metres above the level of the sea, and recorded the following interesting observations: "This excursion yielded very important results in relation to the geological formation of the mountain chain. I found on the high ground extensive remains of a great stream of lava which I could trace down to Lake Tsana. There were craters half filled up and very considerable volcanic cumuli. The upper strata of the mountains consist of crystalline schist, but the lower are composed of the same sandstone formation in which tertiary coal deposits were discovered at Tschelga." *

In an interview reported in the *Egyptian Gazette*, Mr. William MacMillan spoke disparagingly, as it seems to me, of the resources of the country.† I dissent completely from the opinion that Abyssinia has little to offer to commercial enterprise. Its potential wealth is enormous, and if difficulties of transit could be overcome—which means, in this connection, if the Blue Nile could be made navigable—I believe that a vast and most lucrative development of commerce and industry would follow. As the question appears to me to be of great importance, and as I think that Mr. MacMillan's

* "Mittheilungen der Afrikanischen Gesellschaft in Deutschland," vol. iii. p. 27.

† "Asked about the trade prospects in Abyssinia and the projected American commercial mission, Mr. MacMillan said, 'The greater part of the trade of Abyssinia is already in American hands. The principal import is white cotton sheeting, which every Abyssinian wears, all of which comes from the United States. I do not see that there is much possibility of further development in the trade as far as this sheeting—which is everywhere known as Amerikani—is concerned; the supply has already reached its maximum. There may be some possibilities in the case of oil, which at present comes from Russia. As a matter of fact there is no money in the country." (*Egyptian Gazette*, October 21, 1903.)

words may spread a quite erroneous impression, I hope the reader will forgive me for bringing forward the following evidence.

Consul Plowden reported in his "General Survey of Abyssinia," 1852–53 : "Gold and copper exist, and iron is found in great abundance; plains of sulphur and various salts, in the province now occupied by the Taltals, supply all Abyssinia with those commodities: and other wealth may lie hid in that volcanic tract. A search for coal would, elsewhere, be probably successful." *

Dufton visited the district of Tschelga with M. Lejean, the French Consul, who was then taking gifts from the Emperor Napoleon III. to King Theodore. M. Lejean found, when he attempted to make observations in this neighbourhood, that "the ferruginous nature of the rocks destroyed the determinative power of his delicate compass, sometimes, when placed on the ground to a matter of 90 degrees. The presence of iron was further evidenced by the slimy yellow deposits of oxide which some of the mountain rivulets make in their course. Beds of an inferior coal we also found in the plain of Tschelga, laid bare by a small stream which had dug for itself a passage of some forty feet deep. The fact of the presence of coal, not only here but also in many other parts of Abyssinia, seems to point to sources of wealth possessed by this country, which only an enlightened government is required to open out." †

Mr. Vivian considers that the "one insuperable objection" to industrial enterprise in Abyssinia is "the monstrous craft and subtlety of the Abyssinian." He wrote, "You or I might spend ourselves and our treasures

* Quoted in "Abyssinia Described," by J. C. Hotten, p. 129.

† Dufton, "A Journey through Abyssinia," p. 63.

in discovering coal, or copper, or iron, or gold, or emeralds; we might call new industries into being and establish an era of prosperity; but the Abyssinians would take all the profit, and we should be left out in the cold." *

It seems to me, if one may say so without flippancy, that the introduction of a few mineowners from the Rand would soon show the Abyssinian that he is "a child in these matters." A fertile country with an enormous range of climate, capable of well-nigh infinite variety of production, well watered, well wooded, and endowed with gold, iron, and coal is something more than a small mart for American shirtings and Russian oil.

To return to Tschenkar. The soil in this region is very rich, and there is a considerable cultivation of durrha and other grain and of chillies for cayenne pepper. This is the condiment universally used in Abyssinia. It is cooked with or added to every dish, and the natives, including the children, eat it in quantities which scorch the most hardened European gullet. When it is remembered that the Habashes usually eat their meat raw, not even rejecting uncleansed tripe,† and that they are constantly passing from fasts which they observe with superstitious rigour ‡ to an excess of gluttony, that they are immoral from their early years,§ and that drunkenness is an uncensured habit even among the priests, it is surprising that the race has retained its vigour. I am, however, bound to say that the Abyssinian is, normally, hardy and cheery. No doubt the weakly die in infancy or childhood, and

* "Abyssinia," p. 124.

† "Modern Abyssinia," by Augustus Wylde, p. 345.

‡ "Life in Abyssinia," by Mansfield Parkyns, p. 278.

§ Ib. p. 254.

the development of those who survive is greatly helped by the fine air in the upland country.

I made inquiry of our interpreter Johannes as to the system of land tenure here, and was interested to learn that the "Lord of the Manor" leased ground to tenants on the *métayage* system. He himself paid tithe in kind to the Ras. I fancy the landowner takes the lion's share of the crops; and the peasantry are heavily mulcted by the priests and scribes. In addition, it is the rule that all guests of the Ras, when travelling within his fief, must be supplied with food by every village at which they halt. So the farmers have to thrive on what is left.

We were, officially, guests of the Ras. In consequence we found that gifts of *teff*, and poultry, and eggs were brought in without stint. All are cheap commodities in Abyssinia. Augustus Wylde calculated that at Bohoro in the Yejju province, when he was there, fowls were selling at a rate which would have given eight hundred and eighty for the pound sterling, and that the same sum would have purchased eight thousand eight hundred eggs.* The eggs presented to us were nearly all bad, a fact which might be due to a misapprehension about European taste or to a feeling that waste articles might fairly be got rid of as unrequited tribute. The fowls made good eating, but they are very small birds. I do not think the Habashes, at the end of the reckoning, had much reason to regret our presence.

In the evening we received a visit from the priest, a pleasant, fine-looking old man. He wore the turban of his class. We promised to see him at his church early on the following morning and present an offering.

The dry grass had been fired in places around the

* "Modern Abyssinia," p. 369.

village, and the flames showed impressively after nightfall, though their effect was dimmed by the light of the moon, which was nearly full. We had no reason to fear being burned out of our camp, for the stalks had been beaten down all about into a matted mass. But clouds of smoke and showers of blacks from a conflagration close by would have given us more discomfort than we bargained for, and we watched the drift of the fires in the light, varying winds with interest. Luckily, trouble was not for us that night, and we turned in, with all the contentment of the well-fed Briton, and slept in peace.

CHAPTER VII

On the morning of January 12, we fulfilled our promise to the priest of Tschenkar and went to see the church. It has the reputation of being an exceptionally holy place, because the Dervishes made several attempts to burn it down when they raided the village during their incursion after the battle of Gallabat, and their endeavour failed. I cannot tell how the roof fared, for it was made of the ordinary thatch used in Abyssinia when I saw it and looked very inflammable. But it was hardly necessary to attribute the preservation of the rest of the fabric to a miracle. In many cases the entire structure consists of a timber framework with a covering of thatch, and I never heard that these buildings possessed the immunity of Shadrach, Meshach, and Abednego. But the church at Tschenkar is built in part of hard baked mud bricks and in part of stone, and the doorways are made of an extremely tough wood which is found in the country. This had also furnished the beams within the building. Under these circumstances the only marvel seems to be that the Dervishes repeated their attempt to kindle the materials.

The church is round and stands in a circular enclosure, according to the custom in Abyssinia. The wall of the churchyard is of stone, and there are four entrances, facing north, south, east, and west. Cypresses—which

I did not observe in other places—were growing in the enclosure, which, as usual, is just a small grassy expanse. No tombstones or other monuments to the dead are seen in it. The church also has four doors and is divided into two parts. These do not closely correspond to nave and chancel; for the *sanctum sanctorum* of an Abyssinian place of worship is walled in, and makes a third enclosure. Moreover, it does not contain an altar but a representation of the Jewish Ark of the Covenant, which is called a *tabot.* I shall have occasion to allude to this custom and its origin in a later chapter.

The priest showed us round the building. The walls are decorated with the usual extraordinary "sacred subjects," treated in the manner which convention strictly prescribes. St. George and the Dragon and the Virgin Mary are almost invariably depicted, and I was fortunate enough to obtain some clear photographs in the church at Korata,* which exemplify the singular devotional art of the Habashes. At Tschenkar there were some rude carvings of cherubim and designs of the Virgin drawn with burnt wood on the doorways. We were not admitted to the *sanctum sanctorum.*

Outside I obtained a photograph of the theological class, which was composed of one teacher and four pupils. Their library consisted of a single book. I could not, of course, decipher the text, but the volume had the appearance of a very dirty and greasy manuscript missal. Probably it was a copy of the Scriptures, in the Gheez language. There was a quaint thatched belfry in a tree, which looked like a primitive bee-hive or an old bird's-nest.

We had ordered that the baggage donkeys should be loaded and take the road while we were at the church.

* See p. 120.

But we found that confusion always arose if no European was present at the start, and on this occasion, when we had spent some forty minutes with the priest, and then made our way to the track which we were to follow, we beheld no trace of the expedition. When the train did draw in sight we found that one part had been separated from the other, and the two divisions were approaching from opposite sides. It was only a slight *contretemps*, but we lost an hour by it on a day when we had a long march before us. Matters were righted and we moved ahead.

Dupuis and I were bringing up the rear. At half-past eleven we found that the whole expedition had halted on the bank of the river Magetsch, and one of the guides was insisting that we must camp there. He seemed to have no reason for the choice except that there was water in the stream and that the place which we purposed to reach, called Ambo, was, in his opinion, too far off. All the Habashes were of the same mind as the guide, but we did not take their view, and set to work to cross the river.

Here—about a mile and a half from the lake—the current was some twenty feet wide. It was clear and shallow, running over a shingly bed in a ravine. We forded it without difficulty, and moved on towards the camping-ground which we had selected. We arrived at the place shortly after two o'clock.

Our route crossed the two effluents of the lesser Gumara River not far below Wansage. A concise and interesting account of the geology and botany of the river-valley at that place has been given by Dr. Stecker.*

* "The Gumara here flows between high rocks, which consist of porphyry and trachyte, in which calcareous tufa is found embedded. Its banks are adorned with tropical vegetation, and the place derives its

This morning while we were on the road we met the "Sultan of Delgi." The "Sultan," notwithstanding his title, is a subordinate official, the tax-collector of a district. It was this man whose "Deputy" had met us when we arrived on the plateau and raised difficulties. We found the "Sultan" a very affable old man. He shook hands with us, asked if matters went well with the expedition, and if we were in need of anything. This was the more satisfactory as he was then returning from a visit to the feudal lord of the district, Ras Gouksha.

The village of Ambo was destroyed by the Dervishes, and had not been rebuilt. The site of it is on a beautiful little bay of the lake, which at one part has a sandy beach that shelves into the clear water and at another is bordered by a thick fringe of reeds. Snipe swarm in these. There was no cultivated ground at this spot,

characteristic features especially from the wild *Phœnix* with its delicate foliage and the colossal *Musa Ensete ornamentalis* with carmine leaf-panicles, which is called *guma-guma* by the natives to distinguish it from the edible *Musa Ensete.* In many places the Gumara is deep and full of fish; innumerable wild ducks and geese, *scopus umbretta*, etc., enliven the surface of the water, while bright-hued butterflies, the charming *Danais* and equitidœ with their brilliant play of colour, mingled with tiny metallic-tinted nectariniœ (*affinis metallica*), flit around the fragrant *Dodonæa viscosa* and the brilliant blossoms of the Tsana-tree" ("Mittheilungen der Afrikanischen Gesellschaft in Deutschland," vol. iii. 1881, pp. 23, 24). *Phœnix* is the name given to a family of pinnate palms, of which the Egyptian *Phœnix dactylifera*, the finger-date, is an example. *Ensete* is the Abyssinian name of a thick-stemmed, large-leafed banana, which is grown in gardens as an ornamental plant. The thick flower-stalk is used as an article of food by the natives, but the fruit, being small and dry, is useless. In spite of these attractions, Wansage, (see p. 63) according to Stecker, would hardly be a suitable health-resort for European invalids. He wrote, "By night an infernal din prevails, which, mingled with the singing of women and clapping of hands, forms the characteristic music of the Baths. The bull-frogs that live in the Gumara supply a harmonious accompaniment, which, with the noise kept up by the donkeys and beasts of burden, does not cease till long after midnight" (Ib.).

and no timber or "scrub." Tall grass was growing everywhere, and I have no doubt that the soil is extremely fertile.

The tents were pitched, and after lunch we took out our guns to replenish the larder. We made a bag of some half-dozen snipe, and saw an abundance of geese —but they are not very good eating—herons, and other large and small water-fowl. On the way back to camp we put up a flock of crown crane and could have brought down some of them. They are excellent at table; but it seemed a scandal to kill such fine fellows, and we did not shoot. These are noble-looking birds.

In the afternoon some natives passed the camp, who had shot a tree-boa. They were carrying the skin. Without the head, which they had cut off, the snake measured ten and a half feet. As we had avoided camping at a village—a course by which we cheated innumerable vermin—we had no evident claim on any hamlet for free viands. But I found that a fowl and some eggs had been brought into camp, and, to do the Habashes justice, I do not think they grudged the tribute in kind which our presence exacted from them. It is a custom of the country to offer a gift in order to receive one of greater value in return, but the rural folk of the lake country had no such motive—as far as I could see—in furnishing supplies.

We started on the following morning at a quarter to eight. The temperature was then below the freezing-point, and we felt the cold keenly. Our road lay, for some distance, through the tall jungle grass, but when we emerged from it the scenery changed completely. We entered a broken, mountainous district, where the heights rise directly from the lake, and the traveller passes from one surpassingly lovely vista to another.

Many palms grow here, and the banks and dells are covered with wild flowers—a varied, abundant growth excelling even that which we had seen on reaching the lake plateau. Great numbers of butterflies, large and small, familiar and unfamiliar, some of sober hue, others brightly gorgeous, were flitting among the blossoms. The giant hemp abounds here, and the stems, some eighteen feet high, tower above one. At that season the daytime and the evening were alike cloudless. The whole region is fertile and charming.*

It is—for that country—thickly populated. There are many villages, groups of scattered *tokhuls,* and as we advanced we found these yet more numerous. They lie, perhaps, four miles apart on an average. Cattle are seen in all. Very few cases of serious sickness came to my notice hereabout, and the people generally seemed healthy, as well as prosperous. I saw no money in circulation, and all trade appears to be carried on by barter.

As we drew near to our halting-place for the night, those who were leading our column came upon two *tokhuls* full of Habashes. They were about fifteen in number, and, as they carried rifles of an old French pattern, there is very little doubt that they were soldiers. I was in the rear, and when I passed them they were

* This is the district called Lamge. Stecker said of it: "I consider Lamge the prettiest place on Lake Tsana; the ancient, massive *Dokma* trees are all covered by parasitical *Loranthidæ* with rose-red and purple blossoms, and by whole nests of olive-green *Viscum,* and another parasitical plant which is peculiar to this one tree; and they are surrounded by an impenetrable growth of *Cucurbitaceæ* and *Convolvulaceæ,* which in some places form the most romantic piazzas, and in others arbours and formal galleries, and afford the coolest shade. This Lamge is a place for which one may prophesy a brilliant future. Everywhere hereabouts on ancient acacias hang the nests of weaver-birds (*textor alecto*), and, out of curiosity, I counted upon one of them 872 basket-shaped, suspended nests" ("Mittheilungen der Afrikanischen Gesellschaft in Deutschland," vol. iii. 1881, p. 26).

squatting on the ground and staring at those who rode by in a manner that was evidently meant to be grossly insulting. Moreover, they had kindled the grass on both sides of the track, and we were obliged to trot ahead in the midst of smoke and flames. Oddly enough, both the donkeys and mules took the adventure quite quietly, though the fire was close enough to singe their hair and burn their feet. The Abyssinian "regulars" are a class apart, and are unpopular; they are mostly rapacious, hectoring, and ill-conditioned, and are a bane to all from whom they can levy exactions.

Our camping-ground was close to the village of Tschera, which stands on an islet near the lake-shore. It is reached by wading through about fifty yards of water lying over firm sand. Further out is the pretty island of Mitraha, on which we could see houses and a church. My companions took out the boat and their guns and brought back a brace of duck and a green pigeon. Earlier in the day five guinea-fowl had been added to the larder.

We sent two soldiers to buy meat for the men. They purchased five sheep, for which they paid four dollars. This works out at a rate of about one and sixpence per sheep, English money.

We did not intend to resume the march on the following morning, so none of us rose before half-past six, when "the day was aired" a little. The lowest reading of the thermometer at night had been thirty-two degrees. After breakfast, we three Englishmen and Johannes the interpreter started in the boat to visit Mitraha. It is a hilly, tree-clad little island with a population consisting of six families. The people welcomed us on our approach and guided us through a channel between rocks to a good landing-place. They seemed pleased to show us over the ground.* It is a perfectly pretty islet, with quaint thatched cottages among foliage and a ruinous old church, and—for historic interest—it contains the tomb of an ancient king of Abyssinia, John by name. We visited the sepulchre and saw the open coffin, which was full of an agglomeration of bones. If they belonged to the dead monarch, he must have been the boniest man that ever lived, and embarrassed, into the bargain, by possessing two right arms. However, it was an article of belief with everybody on the island, including the priest, that nature had once contrived to stow all that "anatomy" into one human body.

The island is traversed in all directions by narrow tracks marked by trodden leaves, and there is a thick undergrowth of weeds, thistles, and thorns. In this tangle I saw numbers of large spiders' webs, from three to four feet in circumference. The spiders are about the size of a shilling, with a speckled abdomen and legs of enormous length. I saw none of this species on the mainland.

The islanders had poultry but no cattle. We asked them about their food. They said, "We eat durrha when we can get it."

"Don't you eat fish?" we asked. "Surely, there are plently of fish in the lake."

"*No*," they answered emphatically. "After King Johannes was killed by the Dervishes† all the fish disappeared."

It is quite likely that these people are too idle even

* Europeans are not allowed to visit the Island of Dega, "because it is not permitted to any one to set foot on this holy ground, which is only inhabited by anchorite monks and is sacred to St. Stephen" (Stecker, "Mittheilungen der Afrikanischen Gesellschaft in Deutschland," p. 31).

† At the battle of Gallabat in 1889.

to cast a net,* for it is notorious that fish of many kinds and of large size abound in Lake Tsana; but, knowing how superstitious the Habashes are, I would not vouch for it that they did not sincerely believe what they said. Probably they have little intercourse with the mainland and invent their own news.

They took a keen interest in our boat, and asked many questions about it. Their own rafts, called *tankoa*, consist merely of a bundle of papyrus reeds strung together with strips of hippopotamus hide and stiffened by a bamboo in the centre which serves as a kind of keelson. They do not use a sail, but punt or paddle with a pole.

When we had finished the tour of the island Dupuis presented five dollars to the church and community and thereupon received the ceremonial blessing of the priest.†

We returned to camp. During the afternoon I tried my hand at butterfly-hunting, fly-fishing in a sluggish stream near by called the Kumon, and shooting beside the Lake. It was all pleasant enough sport, but I had poor luck. The butterflies that were worth catching were always seen when we were in "marching order," and my net was packed up. When I cast in the Kumon my tackle mostly got foul of the reeds and the weeds,

* I saw no rods or hooks used by natives in Abyssinia. They have nets weighted with pieces of lead. These are simply thrown out to sweep the water at their full spread and slowly pulled in. Then the net is gathered up, hung over the fisher's left arm and cast again.

† Stecker did not visit the Island of Mitraha. He wrote: "On account of a quarrel that arose between my servants and the tatterdemalion clergy of Mitraha I did not land on this parsons' island. The priests wanted to extort money from me by this method: that they would sell me no provisions before I had visited their church, which meant, before I had given them bakshcesh in profusion" ("Mittheilungen der Afrikanischen Gesellschaft in Deutschland," p. 26).

and while tramping along the shore I found nothing to shoot, unless it were a huge hippo who was basking in the sun about three hundred yards out. I should have been *impar congressus Achilli* if we had tried conclusions, so I did nothing to provoke enmity.

In the afternoon I was on medical duty for native patients. The first was a poor little fellow about seven years old, who was covered from head to foot with craw-craw, a kind of aggravated *scabies*. He was brought to me by his father, to whom I gave half a cake of soap, and bade him take the child to the water and wash him. I saw that this was properly done, and then treated the boy's skin with a weak solution of corrosive sublimate and afterwards smeared boric ointment over him. The father was set to wash his son's shama with the remainder of the soap—an office of cleanliness which had been neglected at home in accordance with Ethiopian usage.

My other patient was a man whose upper eyelids were inverted (entropion), as a result of ophthalmia. This malady is very common in the country. I removed the eyelashes with a forceps and applied ointment, but the extent of the injury was already so great that little good was likely to come from my intervention. The Russians maintain a Red Cross Mission at Addis Abbiba,* which has established a hospital on an adequate scale. Here natives receive treatment free of charge.† European medical aid is sorely needed in the country, and I wish that it might be found practicable to support a British medical station in Western Abyssinia. Our popularity and prestige would be increased by this step. An opportunity would be given for valuable observation

* Wylde, "Modern Abyssinia," p. 417.
† Vivian, "Abyssinia," p. 186.

and research, and the benefit to the inhabitants would be incalculable.

Gifts of bread (teff), eggs, and fowls were brought into our camp in plenty, and some liberal Habash sent us a big jar of honey, the comb and all packed in together. It was a welcome and useful present, and served as a luxury in lieu of jam, of which we had a short supply. The size of the jar—it rather suggested the lurking-places of the Forty Thieves—embarrassed us, and we did not wish to carry quantities of honeycomb through Abyssinia. So we consulted Johannes. He told us to heat the jar until the honey just boiled, skim off the wax and pour the clear liquor into bottles. Fortunately we had saved empty bottles, and we filled up all we had; but half the jar remained after that, and had to be returned to the giver. All three of us were interested and busy in melting the honey and straining it through a piece of mosquito curtain, and were splendidly sticky and messy when we left off.

While we were in the kitchen-quarters, we heard a loud grunting and snorting about twenty yards away, which made the ground tremble, and recognized at once that a hippo was in the reeds there. This badly frightened the cook, who slept in the tent nearest that spot. We had to assure him solemnly that the hippo would not carry out a night attack upon him.

We struck our camp next morning (January 15), and marched to the River Reb. The boat was brought into service in crossing it. At one part the stream was about four feet six inches deep, and though this dip in the bed was only a couple of yards wide, it prevented the baggage animals from walking through. The water was shallow over the rest of the ford, so that they could keep their feet. All the mules and donkeys were unloaded while we lunched. The boat was quickly got ready and about half-past two Crawley pulled across with his first cargo. After he had had about an hour of the work, I took my turn. Dupuis was superintending the pitching of the tents and the stowing of the loads on the further side of the river. Then the fun began with bringing the baggage animals over.

There was no great difficulty with the mules. One gave a lead, and then the rest swam across, the boys swimming beside them. But with the donkeys it was a different matter. The men tried coaxing them, leading them, whacking them. It was all no good. The donkeys settled into the firm, resolute stubbornness of their kind, which is invincible, and it seemed as if the expedition would be left in the condition of a wasp that has been cut through the middle—the active half in one place and the "portage and storage" half in another. Sunset began before progress had been made, and the air was quickly becoming colder. Finally we solved the problem by brute force. Five fixed nooses were tied in a long rope, and five donkeys' heads were put in the nooses. Then we, on the further side of the stream, hauled on to the rope. It was a "tug of war," and we moved the donkeys. Then one of us took the rope back in the boat, and the next batch came over by "cable ferry." There was a chance that the donkeys would struggle and strangle or drown themselves, but all were pulled across without casualties. The last three made the journey just after six o'clock. Of course our men had to be in the water with the beasts to steady them, and keep them in place, and all worked with a will. Some were in the stream, off and on, for three or four hours, and the cold was sharp when the sun sank. Unluckily there was no wood near our camping-ground, so they had no fire to cheer them afterwards, and could not dry their wet clothes. The Habashes, when they cross a ford of this kind, roll all their garments into a bundle, which they carry on their heads while they swim or wade. We were all sorry for the mischance which denied the boys warmth after a wetting, and we served out a nip of green chartreuse to those who had been in the water longest.

On the next day, January 16, we had before us the business of fording the river Gumara, and, to avoid a repetition of the trouble with the donkeys, we determined to strike away from the lakeside, so as to reach the stream where we could go over on foot. Dupuis found that this route would bring him close to the Debra Tabor road, and at once resolved to set off, on the following morning, with a light baggage train, to pay a visit of thanks to Ras Gouksha, who was at that place. Crawley was to accompany him, while I stayed in charge of the camp.

It was a short march from our halting-place beside the Reb to the ford by which we were to cross the Gumara, and we did not start till eleven o'clock. The baggage animals were able to have a feed of grass, and they and the boys alike were warmed by the sun before our journey began. All needed a little rest and comfort after the previous day's work.

It is a plain country between the Reb and the Gumara, partly covered at that season by tall, dry grass, and partly by swamp-grass standing in ooze. In the wet months the whole tract is a morass.

Many Habashes assembled to see us start. They were tiresome, but quite friendly. I was much struck by the prevalence of eye disease among them; *trachoma* is very common, and most of the cases which came under my notice were beyond hope of cure. A crowd gathered round me to gaze at my camera, which was an unknown marvel in this land; for Western Abyssinia has seldom been visited by Europeans, and I heard of none who had been in the district since Stecker had passed through it more than twenty years before. Even cigarettes were a novelty. I put one between my lips when I had shut up my camera, and offered another to a young man who was standing close to me. He did not know what it was, and hesitated about taking it. To gratify my curiosity, I pressed it on him, and he held it at arm's length, and looked at it with wonder and suspicion. But when he saw me light mine and smoke issuing from my mouth, he dropped his cigarette as if it were a live shell with a fuse smouldering in it and ran. Crawley gave a cigarette one day to a Habash, who ate it, paper and all, and said he liked it.

We reached the Gumara at half-past two, and pitched our camp beside the ford ; it is an unpleasant place, in the line of a frequented highway. During the afternoon we passed the donkeys in review, and picked out the best for Dupuis's expedition to Debra Tabor. We also made a selection of stores for the journey, which was expected to last five days in all.

CHAPTER VIII

On the way between our camp and the Gumara River we passed many villages. In one I saw a leper in an advanced stage of the disease. Thereupon I made inquiry, and was told that leprosy is very prevalent throughout Abyssinia. Almost all the maladies that work havoc among the people could be gradually checked by adequate medical control under European administration.

On Saturday, January 17, I said good-bye to my two companions, who started for Debra Tabor at nine in the morning. It was the proper course to see Ras Gouksha, and thank him personally for allowing us to travel through his territory, and we desired to obtain from him a letter of introduction and recommendation to Tecla Haimanot, the King of Godjam, whose dominions lie on the other side of the lake. This potentate is named after the most popular, venerated—and apocryphal—saint in the well-filled Abyssinian calendar. I shall refer to his miraculous exploits and experiences in another chapter. They are unequalled by anything to be found in the Golden Legend, and one can only regret that the late "Thomas Ingoldsby" had never heard of him.

The emperor, when he is powerful enough to do so, exercises a suzerainty over the King of Godjam, who served with the Abyssinian forces at the battle of Adowa. But the difficulties of a suzerainty unwillingly accepted, are as manifest in Eastern as they were in Southern Africa, and the country has often been in revolt. For instance, Consul Plowden wrote that in his day the ruler of Abyssinia, who was then the Ras of Begemeder, though a titular emperor was alive,* "had been engaged in the siege of a hill-fort in Godjam now for four years; and another chief in rebellion, after gaining two battles, had pillaged Gondar, and rendered all communication with Godjam circuitous or dangerous." † Menelek's safe-conduct letter is not valid in Tecla Haimanot's dominions, and Dr. Stecker, who was provided with this passport, was refused admission to the country. The German explorer wished to cross the Blue Nile in the neighbourhood of Woreb, but was unable to carry out his plan because the escort officer who had been attached to his party declined to proceed into Tecla Haimanot's jurisdiction. The doctor pointed out, with the logic of his race, that his permit was issued by the Negus Negesti, and that a mandate from the king of kings was binding on the

* Stern gave the following interesting account of the monarch *de jure*:—"Whilst at Gondar we visited various personages of rank and dignity; amongst others *Atzee Johannes*, the Shadow King, and according to Abyssinian annals, the legitimate successor to the throne and lineal descendant of Solomon by the Queen of Sheba's son. He was seated on an *alga* in a dirty little room, with a monk's skull-cap on his head, reading the Psalms. He asked me many questions on geography, and unlike the majority of Abyssinian savants did not believe that beyond Jerusalem the sun never shone, and that only serpents and other venomous reptiles occupied the untenanted land. I was told that whenever *Atzee Johannes* visits King Theodore, the latter stands before him, as an acknowledgment of his title to a crown which he could not defend. He receives an annual pension from the Royal Treasury" ("Wanderings among the Falashes in Abyssinia," pp. 212, 213).

† Quoted in "Abyssinia Described," J. C. Hotten, pp. 122, 123.

A CASE OF LEPROSY. [*See p.* 103.

ruler of Godjam. The escort officer disregarded logic after the manner of his kind, and Dr. Stecker failed to cross the river. He again tried to enter this territory from the north at Wendige, and was again turned back. A third attempt, which promised well, was frustrated by the double dealing of Litsch-Abai, Governor of Wendige, and though the traveller finally received a special permit from King John himself as Tecla Haimanot's suzerain, he was unable to make use of it, except for a few hasty excursions by water from Korata, for he was summoned back to the Abyssinian court before he could accomplish the journey in Godjam which he had planned.*

I took a photograph of my two companions while they were crossing the river Gumara, and then at once returned to the camp and made arrangements to remove it to a less inconvenient site.

I started on my mule with a native guide and an escort of two boys, and in an hour and a half reached a perfect camping-ground about halfway between our last halting-place and Korata. It was a quiet spot on a little plateau, where a spring rose from the rock and ran down a clean, hard channel. This was a welcome change from the muddy water which we had been obliged to use during the last two days. I returned to camp, and after lunch found—for the first time in my life—my reputation embarrassing. Patients poured in, and it occurred to me that if they had had any money and I had wished to establish myself in practice in Abyssinia, I should have done very well by the lakeside. The diseases which I had to treat were leprosy,

* "Mittheilungen der Afrikanischen Gesellschaft in Deutschland," 1881, vol. iii. pp. 26-31.

ophthalmia, tænia, malarial fever, itch, and certain maladies of another order.*

Among my visitors was the parish priest. I supposed he had come to ask alms, and sent a message intended to put off the demand, explaining that I was only the doctor of the party, and that if there was anything which he required, he must return when "the Bey" (Dupuis) came back to camp and tell him about it. To my surprise he sent an answer that he wished to see me and not the Bey, and that he was not seeking alms. That being so, I was very pleased to receive him, and found that he wanted treatment for the itch.

When he had gone I took my rod and tried a fly in the muddy river; but the fish did not rise. So I gave

* Stecker wrote: "Es gehen nach Wansage Kranke aller Art, und da es wenige Abyssinier giebt, die nicht syphilitisch wären, so sieht man meistens nur Patienten, die gegen Lustseuch und ihre Folgen hier Heilung zu finden glauben" ("Mittheilungen," p. 23). As far as my experience goes, this overstates the case a little, but the complete disregard of morality by both sexes—a matter about which the clergy do not trouble themselves—has rendered disease of this kind extremely common. The pest of internal parasites, which is the bane of all classes of Abyssinians, and is almost universal among them, is undoubtedly due to their disgusting habit of eating raw meat. They devour flesh warm and quivering from the newly slaughtered carcase, and Bruce asserted that in his time they habitually cut the meat from the living animal, and continued to do so until it bled to death. This would be an unusual barbarity, and Parkyns saw no instance of it, though he believed Bruce's statement to have been true when he made it" ("Life in Abyssinia," p. 204). Mr. Wylde has given another explanation of what Bruce observed ("Modern Abyssinia, p. 153). But, whatever may be the fact as to this particular, the habit of eating uncooked flesh prevails everywhere in the country. I noticed that the ceremonial practices of the Mussulmans appeared to save them both from tænia and scabies. But, in spite of their ablutions, they are infested, like the rest of the inhabitants, with body lice. Mohammedans are regarded with disdain by the Christian Abyssinians, and the prejudice against them, as I have said, accounts for some of the uncleanliness of the Christians. To wash much or regularly is to behave like a Mussulman.

FORDING THE RIVER GUMARA. [*See p.* 105.

up casting, which brought me nothing but the armache, and went out with my gun. On the swampy ground bordering the lake I shot three of the big geese that are found there in great abundance. They were for the camp larder, not for mine, as I thought the flesh rank and tough. The boys had often asked me to shoot one and give it to them.

On Sunday, January 18, I moved to our new camping-ground. It was a well-wooded spot, grassy and full of wild flowering plants of many kinds. This pleasant place is called Sara; it is six or seven miles distant from the lake. Game was plentiful. On the way I shot a sand-grouse, and, taking my gun in the afternoon, brought down two plovers of a kind which I had not seen before. One, probably the male bird, had excrescences of yellow skin on each side of the beak. There were snipe and quail in abundance, but I did not get within range of any.

During the day a man came to me who had a splinter of wood about two inches long buried in his leg near the shin. To remove it I had to cut down a couple of inches into the flesh. My patient endured the operation without wincing.

Approaching the camp on the way back, I heard tomtoms, women warbling after the manner of the country and the ringing of bells. I supposed for a moment that a stray shot of mine had injured some one and that the neighbourhood was in an uproar about it, and expected to find a demonstration in progress. But on making inquiry I learned that next day the Epiphany was celebrated, when there is a general feast and holiday throughout Abyssinia. The gathering of Habashes was in a hut near the river where they had come to be ready for the opening of the festival. This is the blessing of the water by the priest, and the people are allowed to bathe in it after the ceremony. I did not myself see that any availed themselves of the privilege.

The Habashes, who had assembled near the stream, began their singular devotions at daybreak. Mansfield Parkyns witnessed the whole procedure of one of these commemoration feasts, and his lively account of it is as exact now as when it was written:—

"As we have already stated while speaking of the fasts, the day before the Epiphany is passed by the priests and other devout men in abstinence until sunset. During the afternoon the Holy Sacrament is administered to the priests only, in their churches. After the conclusion of the ceremony they form in procession, and, accompanied by the defterers or scribes, and bearing with them all the church paraphernalia, go down to the neighbouring rivulet. Tents are pitched near its banks, ready to receive them, and there is a store of comestibles of every variety, with, of course, the usual large proportion of beer and honey-mead; the whole of which good things are from the voluntary contributions of the devout of the parish.

"When the much-wished-for sunset has arrived the feasting begins, and it is fearful to behold with what vigour the half-famished divines set to work. There is abundance for them; for the food being begged as a supply for the ark, or tabote, the superstitious people think that they are doing a very godly act in providing vast quantities, while in reality the only result is that the priests make beasts of themselves. The whole night is often passed in alternate prayer, singing, dancing, and drinking. The songs and dances are both of a religious kind: the latter probably taken from the religious dancing of the Israelites, frequently mentioned in the Bible, is merely a peculiar sort of shrugging of the body and stamping with the feet. The end of these devotional orgies is the administration of the Sacrament before sunrise; but it not unfrequently happens that long before that time many of the priests are not in a very fit state to partake of it, disgraceful scenes of drunkenness often disturbing these religious festivals. During the evening of timkat, or the Epiphany, that I passed at Adoua, several of the holy priests were found to have tumbled into the neighbouring brook, Assam, overcome, as charitably disposed persons may have said, by their religious fervour; though some sinful scoffers—myself included, I fear—suggested that liquor might have been the cause of their overthrow.

"After the Sacrament has been distributed among the priests, the chief priest, raising his hands over the stream, blesses it, and then the people bathe in it. Great men, however, and priests, are sprinkled, to obviate the necessity of their mixing, even in such a ceremony, with the vulgar herd. After this the women dance and sing, and the men engage in various sports." *

The "dancing" of the women seemed to me not unlike that which is customary in Egypt and the Soudan. The chief feature of it was the protruding of the chest and chin alternately. The more forcibly this was done the more excellent was the style of the dancer. During the performance they gasp like whirling dervishes.

The Abyssinian women are not kept in seclusion. Indeed, the liberty which they have would be considered license from a European point of view. But perhaps they had been shy of appearing before a band of strangers. Anyhow, I had had no opportunity of forming an opinion about the girls of Ethiopia while we travelled among the villages. They came in numbers

* "Life in Abyssinia," pp. 282, 283.

to the combined church-service and dance which was in progress by the river, and I then found that some were very decidedly good-looking. I took a couple of snap-shots of one leading beauty, and the reader can determine for himself—or herself—if she is fascinating. Tastes differ in these matters; but it is fair to her to say that the charm of an animated, varying expression and of lithe, graceful movements is poorly represented in a photograph.

I took my place among the crowd, not by preference —for the Habash resembles Gargonius more than Rufillus *—but that I might use my camera effectively. Some of the young girls had one-half of the head shaved, and this, I learned, was a sign that they had not yet "come out." It corresponded to short frocks and hair down the back among us. Marriageable maidens had a tonsure—an odd contrast to our employment of the same sign, more especially as it further denotes that the damsel wishes to be mated. As far as I could discern the dances are arranged by the older women with the object of letting the girls be seen to advantage, and I noticed that the tonsures had all been neatly renewed for the occasion. The hair grows naturally in stiff ringlets, and these, springing up vigorously around the shaven circle, make it a very striking feature. The married women wear their hair plaited in rolls, which are saturated with grease.† The gathering broke up about eleven o'clock in the morning and the "congregation" left the bank of the stream. The "recessional" was the tinkling of a bell carried by a miserable-looking little boy, who was dressed in a red gown. He headed the party; then came the priest,

* "Pastillos Rufillus olet, Gargonius hircum." Horace, *Sat.* I. 2.

† *Cf.* p. 42.

INTERESTED IN THE PHOTOGRAPHIC CAMERA. [*See p.* 102.

WOMEN DANCING AT THE FEAST OF THE EPIPHANY. [*See p.* 110.

bearing a cross, and behind him the laity. It was a quaint and interesting sight.

When the camp was quiet again I turned my attention to entomology and went out with a butterfly-net and a tin box to hold captures. Behind me was a boy carrying my gun—a sort of Friday to my Robinson Crusoe. We walked down the stream, and I saw many green swallow-tail butterflies, but they were much too quick to be caught by my pursuit. Presently I noticed that they were attracted by a certain red mossy growth in the water, so I waited near a patch of this. A butterfly hovered over it, and I made a dash for the specimen, and got it, but damaged it in the process. Then I laid it, in as life-like an attitude as I could give it, on the moss, and by this means secured four fine specimens which came to see what my decoy had found.

In the afternoon I wandered up stream and saw one snipe, which I shot as part of the menu for dinner. Then I left the course of the river, and walked towards some trees which seemed to form a detached clump. But as I drew nearer and fresh ground became visible, I found that they were part of a fine wood. When I was close to the edge of it, four quail rose at my feet. I tried to hit one, but was not sharp enough. At the noise of the discharge a large bush-buck sprang up in alarm and scampered off among the shrubs. The trees were chiefly of the cypress family, and, following a footpath, I discovered the explanation of this; for I came to an old ruinous church, and the whole grove had spread gradually from its enclosure. Beyond the church there was open ground, and a village lay at a little distance. The scene was sunny, pleasant, and almost idyllic. I retraced my steps and completed my bill of fare by adding a pigeon to the snipe.

The next day (January 20) I was medical officer to the country-side again until half-past ten. One man had a half-inch splinter of wood buried in his leg. He, like my former patient in a similar plight, was very plucky and let me cut down on it and take it out. When I had dressed the wound, he showed the zeal of his gratitude by throwing himself upon the ground and trying to kiss my boots. Several of my "cases" sent me offerings of eggs. About two dozen came in altogether, a couple from one, three from another, and so on. I was gratified, because these voluntary gifts, though they were small, were tokens of good-will. Ordinarily the escort-officer* whom Ras Gouksha, according to the custom of the country, had told off to attend us, and whom we called "the tout," extorted provisions for us from the villagers—bread, milk, eggs, honey, grain, and the like, we being entitled to levy this tribute as the official guests of the Ras. The "tout" had gone with my companions to Debra Tabor, and it was pleasant to find that, in his absence, I received "benevolences" of another kind. This man professed to be our guide for the journey round the lake, but we found that he knew nothing at all about the roads and tracks. I may add that the eggs spontaneously given were all new-laid, while those exacted most emphatically were not.

After lunch I went out with my gun, and shot, besides a brace of pigeons, a pair of fly-catchers of a very pretty kind. Their plumage seemed to be worth preserving and I skinned them in the afternoon. Later

* "This official is called a '*balderabba*,' and is intrusted with a traveller's safety" (Stern, "Wanderings among the Falashes," p. 96), and "becomes a sort of agent, and expects you to acknowledge by presents, any service he may render you, such as assisting you out of difficulties in which you may be involved" (Parkyns, "Life in Abyssinia," p. 72)

I had a complimentary visit from the priest. We exchanged our respects in silence, by pantomime. He bowed and displayed his deference ceremoniously in dumb-show, and I bowed in return, and then the formalities ended. After this I talked to him by the aid of Johannes. This priest was a leper suffering from the form of the malady called "nerve leprosy." He, poor fellow, had not known the fact until I told him. I almost regretted that I had unwittingly ended his ignorance. He showed the greatest astonishment and interest while he listened to an account of the methods employed in the treatment of the disease in Europe.

I was up early on the morning of January 21, and set to work to put a coat of paint on the boat, but patients soon found me out and changed my occupation. One case interested me much. It was that of an old lady, a *dear* old lady, who told me that she was a hundred years old! Why she reversed feminine tradition by saying that, I cannot tell; upon closer observation I judged her age to be about fifty-five, allowing for the earlier senescence of Oriental people. She had heard of me in Tigre, eighty-five miles distant, and had come all the way to Sara that I might examine her broken wrist. The fracture was a fortnight old, and I found that the bone had been set very sensibly, and bound up with a ligature made of strips of cane strung together upon a piece of leather. This made an excellent splint. The injury was in a favourable position, and the *soi-disant* centenarian had full use of her fingers. I set up the fracture just as it had been bound before, only with cotton-wool, a couple of bits of cardboard and a neat new bandage. I thought that, as the patient had made such a journey to consult me, she deserved that much treatment, and she would have been sorely disappointed if she had returned to Tigre with no outward and visible sign of the European doctor's attention to show to her friends. As it was, she was pleased and grateful, and took the homeward road forthwith in all contentment.

About four o'clock in the afternoon my companions returned from Debra Tabor. They had been received three times "in audience" by the Ras, but I did not gather that any very deep impression had been made upon them. They were, in fact, relieved to think that no more visits of ceremony were in prospect. To the grief of all, one of our soldiers had died on the journey. He was seized with fever the day after he started, rapidly became worse, had to be left in a hut some six miles out of Debra Tabor, and succumbed during the evening of the second day of his illness. From a description of the symptoms. I judged that his death was caused by severe intestinal irritation, probably due to ptomaine poisoning. This, in all probability, resulted from eating putrescent meat, a practice which we could not suppress among our men, though it was particularly likely to be fatal in such a climate as that of Abyssinia.

My companions brought with them some guinea-fowl and a sheep, and we bought an ox in camp, which was to be killed the next day at Korata. The cost of it was seven dollars, or about twelve shillings and six pence, the vendor reserving the hide for his own purposes as a condition of the bargain. That night a hyena was prowling round my tent. I was not sure what manner of beast it was until Crawley enlightened me next day. It certainly would not be safe for a traveller to sleep in the open in places haunted by these creatures.

On the morning of January 22, after some delay in starting, we made a steady jog-trot journey, through dried grass in a well-timbered country, down the gentle

THE 'CANDELABRA EUPHORBIA' ON THE EDGE OF A DRY WATER-COURSE. [See p. 66.

BETWEEN SARA AND KORATA. [See p. 114.

slope that trends away from the plateau. The track which we followed was rocky for the most part, and lay in cultivable land. We saw enclosed fields here and there and numerous villages—one in every four miles or so. There was no mishap, except when the donkeys stuck in the mud at the bottom of the watercourses, or lay down in it. And then it was dirty work to get them on, or up. Just before one o'clock we arrived in Korata Bay and pitched our camp there.

After lunch my companions started on muleback with their guns. They were to be guided by a native boy about eighteen years old, whom our interpreter Johannes had engaged in the town, and who said that he knew where game was to be found. Crawley took the young fellow's staff in one hand, and in the other held out a Maria-Theresa dollar, and made him understand by signs that if the sport was good he would get the money, but if no game was seen he would have a cudgelling. The humour of this tickled his fancy very much, and he was convulsed with laughter over it. It was the first time I had seen an Abyssinian enjoy a hearty laugh.

Korata stands upon a promontory. When I had held a "doctor's parade," I took my gun and walked round the point of this little headland. There appeared to be no birds to shoot, and I had made up my mind to turn back empty-handed, when I saw a couple of fat little ducks on a rock in the lake. I shot them when they rose, and, knowing they were dead, sat down to wait till the waves brought them within reach. The noise of the firing roused the curiosity of two Abyssinians, who came up to me and bowed. Then one very obligingly walked into the water and fetched out the birds. After this both accompanied me, the first carrying the ducks, the other bearing my gun. I understood that they meant to show me the way over the brow of the hill and through Korata to our camp.

We left the sandy beach, and walked up a rocky track full of loose stones into the town. It is merely a collection of tokhuls scattered in groups of six or seven among untended coffee plantations. These are not separated by hedges or fences, and perhaps I might best convey an idea of their appearance to an English reader by comparing them to a series of neglected orchards without traceable boundaries. Indeed, I should have supposed that the coffee bushes, which attain a most remarkable size, grew wild here, if it were not for a statement in Consul Plowden's report. He wrote: "The more temperate provinces would probably be favourable to the growth of pepper, spices, and coffee, which last has already been planted at Korata." * Evidently the soil is admirably suited for the raising of this produce to advantage; the coffee is of an excellent quality, and though the natives bestow no trouble and spend no money on the industry, there is even now a steady export trade in the berries from this district. The principal market for it is in the Soudan, where the demand is likely to be larger as the development of Upper Egypt proceeds.

We traversed the main thoroughfare on the way back to camp. It is so narrow that two persons cannot pass abreast. On either side is a bank overgrown with nettles and other weeds. I saw very few people; the life of the place seemed to be stagnant, and here, as elsewhere in Abyssinia, there was very little money in circulation. Judging roughly, I should estimate that Korata contains a population of some two thousand, and that three-quarters of these are Soudanese slaves. It is

* Quoted in "Abyssinia Described," p. 129.

the custom among the Habashes that each man has a boy-slave, and each married woman a girl-slave. The attitude of mind of the Christian Abyssinians towards these wretched people is well shown by the following slight incident—as I was walking through Korata with the two Habashes who had volunteered to guide me, we passed a couple of slave-boys. I said, "Soudanis," and immediately my companions spat upon the ground to signify their contempt.

When we reached camp I gave my new acquaintances a drink of chartreuse in token of my gratitude, and then they squatted in front of my tent, and seemed greatly interested in watching me while I dressed a wound on the leg of one of the donkey-boys. They were doing no harm, but "the tout" bustled up in his most consequential mood to hurry them off, and I think he hit one of them. Anyhow, there was some disturbance, and the other Habashes made common cause. The corporal came to report to me that the escort officer was drunk, and had struck a man. Thereupon I summoned "the tout" and rated him, and ordered him to go and apologize to those visitors of mine, who had done me a service. Matters straightened themselves out, for the two men came to me half an hour later to say good-bye, and they appeared to be perfectly happy and contented. They "bowed and scraped" in the usual Abyssinian manner before they took their departure. "The tout" made some representation to Dupuis when he returned to camp; so I explained that the fellow was tipsy and troublesome, and he was sent about his business. I gathered that he asked leave to go, and hoped that he meant to quit us altogether, but found afterwards that he had only sought permission to spend the night in the town—over more tedj, of course. Dupuis paid no heed to what he was saying, so off he went, but came back in an unruffled mood, and full of complacency, on the following morning. I think Johannes gave us the explanation of his conduct—that he had orders from the Ras to make sure that we three Europeans should have as little intercourse as possible with the Habashes.

That night the noise in our camp disturbed a hippo, whose resting-place in the reeds was close by. I heard him snorting loudly and angrily. He had a prescriptive right to quiet sleep in that place, and I felt that we were intruders upon the ancient order. The old-world life seemed to have a *raison d'être* superior to that of a modern expedition, and though the hippo kept me awake, I sympathized with his indignation.

He and his kind were prominent figures in the scene next morning (January 23). We counted five that were in sight together, and four were less than two hundred yards from the shore.

After breakfast we visited the celebrated church of Korata. I am not sure that it is the only one in the place, but its reputation makes it *the* church of Korata. We were told that a certain Frenchman had travelled from France expressly to see it; if he did, it will be owned that he was an exceptional Gaul, for the length and discomfort of his journey were out of all proportion to the satisfaction to be obtained at the end of it. The building—as my photograph of the exterior shows—is circular, with thatched walls, and in point of structure differs only in size from the village churches. It stands in a compound, surrounded by a rough stone wall. The entrances are two lich-gates, surmounted by thatched porticoes.

The church owes its especial sanctity to the patronage of a female saint. Her chief exploit has been described

THE CHURCH AT KORATA CONTAINING THE FRESCOES. [*See p.* 118.

THE PORTUGUESE BRIDGE OVER THE BLUE NILE. [*See p.* 147.

From a Photograph by Mr. C. E. Dupuis.

by Mr. Hormuzd Rassam, and as the passage in which he recounts it is full of the "atmosphere" of the country, I quote it in full:—"In accordance with orders from the King (Theodore) the two principal merchants of the place (Korata) Aito Kasa and Aito Wandé—the latter, as mentioned above, the person on whom Theodore had bestowed Lij Tasamma's sister, one of his favourite concubines—came out in a wherry to meet us about a mile from the shore, clad in gorgeous habiliments. After an interchange of civilities, we followed them to a small plateau which rises above the lake about two miles to the south of the town, where upwards of a hundred ecclesiastics from the neighbourhood were assembled to welcome us, his Majesty having directed the clergy to receive me with royal honours, and to take care that their persons and garments were scrupulously clean. They were in full canonicals, and greeted us with prayers and psalmody, conducting us with the same accompaniments to a tent which had been erected for the occasion, a few yards from our landing-place. The glitter of the pageant was heightened by a grand display of crosses, croziers, mitres, church umbrellas, David's harps and censers, which were born aloft in the procession. After spending a short time witnessing their religious dances and listening to their discordant chants, the benediction was pronounced, and the Lord's Prayer wound up the service, when we accompanied the merchants to their dwellings on mules provided for our use. Korata rejoices in a patron saint of great celebrity—a native Joan of Arc, in fact—called Waldt-Maryam. She was a resident nun when a formidable Galla Chief, who had overrun a great part of the country, appeared before Korata. She encouraged the townsmen to defend the place, and their valour combined with her powerful intercessions availed to withstand the repeated assaults of the infidel hosts, who were eventually obliged to retreat with great loss. Since that time the town has been placed under her special patronage, and to show their veneration for the locality Abyssinians generally dismount on approaching it, and walk on foot through the streets." *

It is curious to reflect that the name of this remote little township in Western Abyssinia was once very familiar to British ears, and it would scarcely be straining an accepted phrase to say that the eyes of the nation were turned towards this place; for while the British mission was here and King Theodore was encamped on the peninsula of Zegi, on the opposite shore of the lake, the negotiations about "the captives" reached and passed the crucial stage in the early months of 1866. The end of the story, as the whole world knows, was the suicide of the King at Magdala. It is difficult even now, in spite of all that has been written on the subject, to judge how far he was mad and how far sane, how far a capricious, clever, cruel savage, and how far a man of good parts and fine character marred by circumstances.

In the church of Korata we first saw Abyssinian frescoes. It struck me that they had an unmistakable resemblance to the Persian style, and, in some instances, in spite of their crudity, the depiction of the faces is good. The "exposure" photographs which I obtained show the types of the subjects represented. Saint George, who is the patron of Abyssinia as he is of England, is always a prominent figure. The holy people who, by established usage, are admitted to the "goodly fellowship" on the walls include certain worthies of the Apocrypha. The picture which struck

* "Narrative of the British Mission to Theodore, King of Abyssinia," by Hormuzd Rassam, F.R.G.S. Murray, 1896. Vol. i. pp. 317, 318.

FRESCOES IN THE CHURCH AT KORATA.

[*See p.* 120.

me most showed the Virgin and Child crossing the Red Sea—probably during the flight to Egypt—in an Abyssinian rush-boat and feeding hippopotami with bread as they went. They are seated in the stern and are rowed by a Habash.

The church of Korata has a large staff of priests and *defterers* or scribes.* To benefit fully by the sanctity of the place, they sleep in the porches, on the beds which are called *angareebs* in the Soudan. A theological school is attached to the holy building, and we saw five or six of the students. One of them was suffering from the disease which Stecker found to be so distressingly prevalent in Abyssinia. The Chief Priest received us very graciously. He was much interested when we named the saints depicted in the frescoes, and appeared suddenly to realize that we were Christians. He seemed to have the impression that an English official's training comprised instruction in all handicrafts; for he put into my friend Dupuis's hands the separate parts, including the mainspring and all the wheels, of an old silver timepiece, and signified through the interpreter to my astonished companion that he was desired to put them together again and return the watch in working order. Dupuis explained that he was not a "practical jeweller and repairer." The agreeable old dignitary, whose name was John, received an offering of ten dollars with great gratitude.

Later in the day he sent me as a patient one of his students, who was suffering from ophthalmia. This man brought a citron for each of us three Europeans; why, I cannot guess, for this pulpy kind of lemon is not a useful gift, nor a delight for a *gourmet*. The fruit may have had some symbolical meaning, but, if so, we

* See p. 223.

remained in ignorance of it. I saw many other patients. Nearly all suffered from trivial ailments. A woman and her husband, who both appeared to be in sound health, kept trying to tell me some long tale or other, hovered about near me, and drew my attention at every turn. Finally they induced a soldier to interpret for them, and besought me to attend to the woman's case. When I examined her, I found that she had come to consult me about the traces of the scar of an abscess which had healed, as far as I could judge, some twenty years before! This blemish on her "beauty" was about the size of a shilling.

In the afternoon we went shooting. The boy—Zody by name—who had been so tickled by my friend's dumb-show on the previous day, and had earned his dollar by taking my comrades where guinea-fowl were plentiful, again attached himself to our party. He guided us well, and we made a good bag of quail, guinea-fowl, and partridges.

When we returned we found that we had received a present from the lady of the Ras in whose dominions we were. It consisted of loaves of "teff" bread, ten fowls, sixty eggs and a jar of honey. All this we distributed among the men. The Ras's wife was owner of the soil hereabouts; it had been given to her as dowry, and in such a case a woman has practically absolute proprietorship, and the taxes are collected for her benefit. But her authority does not extend to the selection of officials; that power remains in the hands of the Ras. As the Mussulmans had killed during the day the ox which we had bought for twelve shillings and sixpence, the men had plenty of good fare before them, and our own larder was not badly stocked. Vultures of enormous size gathered round the spot where the animal was slaughtered, and fought over the offal that was left on the ground. Numbers of carrion crows had flocked to the feast, and had to be spectators for the most part. But they slipped in among the big combatants now and then, and filched away a morsel adroitly.

I have given rather fully my impressions of Korata, which is a typical township of Western Abyssinia. It may interest the reader if I supplement it by the brief note written by Stecker, who sojourned in the place more than a fortnight. He wrote—

"On April 1" (1881) "we reached Korata, the most important, most charmingly situated and largest town on Lake Tsana. Korata has often been visited by Europeans before, most recently by Piaggia, who stayed here about a year and a half and occupied himself with ornithology. Korata is famous for the first-rate excellence of its coffee, which, as I was able to satisfy myself, flourishes here exceptionally. The place is almost the most important market on the Tsana, but at the moment does not contain more than eight hundred to one thousand inhabitants against three thousand in Theodore's time, as very many have either migrated from the town or died of fever. At an earlier time the Mohammedans were numerous here, but most of them emigrated to Gallabat after the order issued to them by King John that they should one and all embrace Christianity. Only a few families abandoned Islamism and adopted the Coptic faith. There are also three Jewish families living here. Korata is divided into ten districts whose names are Dengelteffa, Tukuwodeb, Margeza, Kulomalfia, Siet biet Negus, Guaguata, Guwi, Adisamba, Vof tschogevia, Gusudur, and Slam biet (or Slam modeb), which last is at present entirely deserted. Korata is the abode of an exceptionally numerous priesthood." Dr. Stecker formed friendships among the laity in the place, but "by no means lived in the best harmony with the pretentious clergy." *

* "Mittheilungen der Afrikanischen Gesellschaft in Deutschland," 1881, vol. iii. pp. 24, 25.

CHAPTER IX

We remained two days at Korata in order to replenish our stock of grain and other provisions. During this time I watched the birds of prey as closely as I could. They are of the same species as those found in the Soudan, and my rough and brief observations, so far as they went, altogether confirmed the interesting remarks of Sir S. Baker on the subject. The passage illustrates so well the brilliant combination in that explorer of keen sportsmanship, and rare ability in the study of natural history that I need offer no apology for quoting it:—

"A question has been frequently discussed whether the vulture is directed to his prey by the sense of smell, or by keenness of vision; I have paid much attention to their habits, and, although there can be no question that their power of scent is great, I feel convinced that all birds of prey are attracted to their food principally by their acuteness of sight. If a vulture were blind it would starve; but were the nostrils plugged up with some foreign substance to destroy its power of smell, it would not materially interfere with its usual mode of hunting. Scent is always stronger near the surface of the ground: thus hyenas, lions, and other beasts of prey will scent a carcase from a great distance, provided they are to leeward; but the same animals would be unaware of the presence of the body if they were but a short distance to windward.

"If birds of prey trusted to their nostrils, they would keep as near the ground as possible, like the carrion crow, which I believe is the exception that proves the rule. It is an astonishing sight to witness the sudden arrival of vultures at the death of an animal, when a few moments before not a bird has been in sight in the cloudless sky. I have frequently lain down beneath a bush after having shot an animal, to watch the arrival of the various species of birds in regular succession; they invariably appear in the following order:—

"No. 1, the black and white crow: this knowing individual is most industrious is seeking for his food, and is generally to be seen either perched upon rocks or upon trees; I believe he trusts much to his sense of smell, as he is never far from the ground; at the same time he keeps a vigilant look-out with a very sharp pair of eyes.

"No. 2 is the common buzzard; this bird, so well known for its extreme daring, is omnipresent, and trusts generally to sight, as it will stoop at a piece of red cloth in mistake for flesh; thus proving that it depends more upon vision than smell.

"No. 3 is the red-faced small vulture.

"No. 4 is the large bare-throated vulture.

"No. 5 the marabou stork, sometimes accompanied by the adjutant.

"When employed in watching the habits of these birds it is interesting to make the experiment of concealing a dead animal beneath a dense bush. This I have frequently done; in which case the vultures never find it unless they have witnessed its death; if so, they will already have pounced in their descent while you have been engaged in concealing the body. They will then upon near approach discover it by the smell. But, if an animal is killed in thick grass, eight or ten feet high, the vultures will seldom discover it. I have frequently known the bodies of large animals, such as elephants and buffaloes, to lie for days beneath the shade of the dense nabbuk bushes, unattended by a single vulture; whereas, if visible, they would have been visited by these birds in thousands.

"Vultures and the marabou stork fly at enormous altitudes. I believe that every species keeps to its own particular elevation, and that the atmosphere contains regular strata of birds of prey, who, invisible to the human eye at that enormous height, are constantly resting upon their wide-spread wings, and, soaring in circles, watching with telescopic sight the world beneath. At that great elevation they are in an exceedingly cool temperature, therefore they require no water; but some birds that make long flights over arid deserts, such as the marabou stork and the bustard, are provided with water-sacs; the former in an external bag a little below the throat, the latter in an internal sac, both of which carry a large supply. As the birds of prey that I have enumerated invariably appear at a carcase in their regular succession, I can only suggest that they travel from different distances or altitudes. Thus, the marabou stork would be farthest from the earth; the large bare-necked vulture would be the next below him, followed by the red-faced vulture, the buzzard, and the crow that is generally about the surface. From their immense elevation, the birds of prey possess an extraordinary field of vision; and, although they are invisible from the earth, there can be no doubt that they are perpetually hunting in circles within sight of each other. Thus, should one bird discover some object upon the surface of the earth below, his sudden pounce would be at once observed and imitated by every vulture in succession. Should one vulture nearest the earth perceive a body, or even should he notice the buzzards collecting at a given point, he would at once become aware of a prey; his rush towards the spot would act like a telegraphic signal to others, that would be rapidly communicated to every vulture at successive airy stations.

"If an animal be skinned, the red surface will attract the vultures in an instant; this proves that their sight, and not their scent, has been attracted by an object that suggests blood. I have frequently watched them when I have shot an animal, and my people have commenced the process of skinning. At first, not a bird has been in sight, as I have lain on my back and gazed into the spotless blue sky; but hardly has the skin been half withdrawn, than specks have appeared in the heavens, rapidly increasing. 'Caw, caw,' has been heard several times from the neighbouring bushes; the buzzards have swept down close to my people, and have snatched a morsel of clotted blood from the ground. The specks have increased to winged creatures, at the great height resembling flies, when presently a rushing sound behind me, like a whirlwind, has been followed by the pounce of a red-faced vulture, that has fallen from the heavens in haste with closed wings to the bloody feast, followed quickly by many of his brethren. The sky has become alive with black specks in the far distant blue, with wings hurrying from all quarters. At length a coronet of steady soaring vultures forms a wide circle far above, as they hesitate to descend, but continue to revolve around the object of attraction. The great bare-necked vulture suddenly appears. The animal has been skinned, and the required flesh secured by the men; we withdraw a hundred paces from the scene. A general rush and descent takes place; hundreds of hungry beaks are tearing at the offal. The great bare-necked vulture claims respect among the crowd; but another form has appeared in the blue sky, and rapidly descends. A pair of long, ungainly legs, hanging down beneath the enormous wings, now touch the ground, and Abou Seen (father of the teeth or beak, the Arab name for the Marabou) has arrived, and he stalks proudly towards the crowds, pecking his way with his long bill through the struggling vultures, and swallowing the lion's share of the repast. Abou Seen, last but not least, had arrived from the highest region, while others had the advantage of the start. This bird is very numerous through the Nile tributaries of Abyssinia, and may generally be seen perched upon the rocks of the water-side, watching for small fish, or any reptile that may chance to come within his reach. The well-known feathers are situated in a plume beneath the tail." *

On the night of January 23, the thermometer sank to 29° F. just before sunrise, so no one turned out with alacrity, and we were later than usual in starting.

I was surprised to find that the boy Zody sought to attach himself to the expedition. While we were preparing to leave Korata he came to us and said, "You are good people and I want to go with you." We raised no objection, and he joined our party. A little while afterwards he declared his intention of accompanying us to the Soudan. Thereupon he returned to Korata and "realized his capital" by selling the cow which he possessed. Then he overtook the expedition, and served

* "The Nile Tributaries of Abyssinia," pp. 492–496.

throughout the journey to Gallabat, receiving the same pay as the other boys. He was smart, willing, and useful, soon learned from his comrades what our requirements were, and acted both as indoor and outdoor servant. He had no Arabic, and all intercourse between us was by signs, which he was extremely quick to understand. The motive that prompted him to render us service was undoubtedly his devoted loyalty to Crawley.

Most travellers have given the Habashes a thoroughly bad character. It is said that they are vain, greedy, treacherous, and cowardly. Though we Europeans had but few unpleasant experiences of them while we were in the country, I cannot say that I saw anything which would give me authority to dispute the conclusions of writers who knew them better from longer observation, and it is the more pleasant on this account to record an instance of an Abyssinian's fidelity and cheerful good-will.

We marched through the familiar dry grass for about half an hour after we left Korata, and then reached the river Gelda at a place where, according to the "tout," it was fordable. This is a muddy, rapid stream, flowing between earthy banks. Its bed here was soft and yielding, and there was not the least possibility of getting the beasts and baggage across. In fact, our guide had misled us. It was a most annoying waste of time. While we were shooting on the previous day we had seen a fording-place higher up the stream, and now struck towards it in spite of the "tout's" assertion that the water there would be up to the men's necks.

When we reached the spot, we found a shallow, fairly fast current flowing over a rocky bed. The guide was entirely unabashed by the fact that he had misdirected us and lied to us. After this we ceased to consult natives and planned our course according to Stecker's map, and in this way managed the matter well enough. There are some small inaccuracies in the doctor's work, but it is entirely sufficient for all practical purposes and a worthy record of the German traveller's high attainments and untiring patience.

The ledges of rock which form the bed of the river at the upper ford are smooth and round, and in consequence very slippery. We had a busy time getting the donkeys over. Men had to be posted in the water to steady them as they crossed, but the real trouble was at the further bank. It was high and steep, and the way up was by a narrow cutting with a hard surface. The first wet donkey that passed along this made it as insecure for foothold as a slide, and the next donkey struggled and scrambled and fell back into the water. We had to tell off boys to shovel relays of earth on to the slope and then others grabbed the donkey's load, its ear, its tail, or anything they could get a grip of, and shoved it up bodily. Another detachment was on duty on the top of the bank to prevent the animals from straying.

Beyond the river lies a steep hill about five hundred feet high. The ascent is by a track of the usual kind, about a foot and a half wide. The way down from the summit is very precipitous and was full of loose stones. These the donkeys often dislodged, and many rolled along and struck the animals that were moving in front, but we had no trouble with them.

We passed a village which is called Selselima, consisting of the usual tokhuls grouped about a round thatched church. Here I saw an example of Abyssinian methods of administration. A lad of about eighteen was standing in a field, watching us pass. Ras Gouksha's man, the "tout," went up to him, grabbed him by his gown, and began to question him. Soon another Abyssinian approached him, grasped the boy's *shama* in like manner, and also interrogated him. I inquired what they were asking, and was told they were seeking directions to guide them by the nearest way to the source of the Blue Nile, which was then some four miles distant. Presently without apparent reason, both our fellows began to belabour the lad with their staves. I rode up and ordered them sharply to desist, and explain to me what they were about. Thereupon they made off, and one of our escort told me they had been beating the boy so as to force him to act as their guide—without pay, of course—and accompany them to the river, though he had already given them full directions for the road.

The end of the descent brought us to the lake shore again, and we camped on the further bank of a little stream of clear water running in a rocky bed. It is called the Tschembolo. This is near to the village of Woreb, and is only two miles from the point where the Blue Nile flows from Lake Tsana. We could not proceed further that day, for the march had been a long one, and it was late already. So we halted upon this spot by the brook, as it was a very suitable camping-ground with a clean drinking supply, and determined to move on the morrow to a more central position, if we could find one, for the survey work which Crawley and Dupuis were to undertake.

Patients had followed me from Korata. They were people who had reached our camp there in the morning, after I had packed up the medicine-chest and mounted my mule. Some half a dozen had come over to Woreb after us, and were now waiting to see me. But I was dead tired that evening and refused to attend to them.

OUR GUIDE BETWEEN KORATA AND WOREB. [*See p.* 132.

There were no serious cases, and I told those who wished to have advice to return at nine o'clock the next morning. If I had made myself more accessible than I was, I should have been mobbed not only by the sick in Abyssinia, but by a host of people who fancied they had ailments.

Just before dinner there was a commotion close to our tents, and we found that one of the soldiers of our escort had tried to stab Walda Mariam,* the assistant of Johannes, with whom he had come to meet us at Gallabat. We immediately disarmed the soldier, and then discovered that both had been to a village near by to get *tedj*, and were drunk. So we disarmed the deputy-interpreter also, and I was sorry to see him in disgrace, for he had always been willing, hard-working, and cheery on the journey. We placed the soldier under arrest, and warned him that if he were found brawling again he would be flogged. Then, order being restored, we went to dinner.

In this part of Abyssinia the mountain air sharpened our appetites, and I must own that we were sturdy trenchermen. Each would tackle a whole guinea-fowl or duck for lunch, and be ready for another in the evening. We had puddings "of sorts" too, and yet, like Oliver Twist, we "asked for more."

On the morning of January 26, the quarrelsome soldier, now sober and crestfallen, was brought up for a formal reprimand. He was told that his case would be reported at Gallabat, and that he would be punished there according to military law. Then his arms were returned to him. Walda Mariam also attended, looking sheepish and sorry for himself, and his weapons too were restored to him.

* Cf. such a Spanish name for a man as José *Maria*.

After this the medicine-chest was unloaded, and I redeemed my promise to patients. One family—father, mother, and two children—had come from a distance to consult me at Korata. They had arrived too late, and had followed us to Woreb. The father was suffering from necrosis of the lower jaw, and I had to draw three teeth for him. The mother had come to hear what instructions were given about the children—a little boy and a little girl. They were suffering from strumous, tuberculous glands of the neck. I had as "hospital orderly" a soldier who understood Arabic and Amharic, and could therefore put me *en rapport* with my patients. After an examination of the children, I concluded that the right thing would be an operation with the knife. I told this to the woman, and the soldier interpreted what I said word for word. As soon as the little girl heard the word "knife," she gave vent to the loudest, most piercing scream that she could utter, and fairly flew from the spot. The noise created a sensation among the Habashes and boys, who seemed to think that I was killing a patient. It would have been awkward in any case to operate in camp, and on this occasion I had no opportunity. Nothing short of brute force would have brought the little girl back, and I think the word "knife" conveyed to her mother and brother the notion of the butcher's rather than the surgeon's implement.

My companions had mounted their mules and started on an exploring journey, intending to select a good site as a camping-ground close to the river. But by the time they returned they had decided that our present position could not be bettered, as the river-banks were low ground covered with papyrus swamps, and any halting-place there would be very unhealthy. They had also found that all the points from which they wished to make observations were easy of access from the spot that we had chosen.

I had a rather sharp touch of fever in the afternoon, and was obliged to give up work and lie down. These attacks are unavoidable at times in that country. They make one feel extremely helpless and depressed while they last, but soon yield to quinine and a little rest.

The next morning, January 26, I was better, but felt shaky when I got on my mule and started with my companions. At a short distance from the camp we left the animals in a hollow and climbed a steep hill about five hundred feet high. It was covered with tall grass and the soil was stony, so that I was soon out of breath. But when we reached the summit a lovely scene lay in front of us. The Blue Nile, winding away from its outlet in the lake, was bright in the sunshine. The green banks on either side were delightful to the eyes, and here and there in the stream were little islands, vivid in colour, where the papyrus grew to a height of twelve or fourteen feet. There were many dark dots in the water—heads of hippopotami, which swarm in these upper reaches of the Blue Nile. Mountains rise above the river-valley on both sides, and the stream curves among the spurs at their base, till it is lost to sight. The view from the high land is far-reaching, and the clearness of the air makes even distant outlines very distinct. I can scarcely imagine a scene at once so charming, tranquil, and impressive as this prospect of lake, river, and mountains. It is impossible to convey in words the effect of the simple, strong colouring—the blue lake, the flashing stream, the verdurous islands and swamps, the cloudless, lustrous sky, the chromes and grays and purplish shadows among the ridges that sweep upward and recede from the valley in splendid lines. And upon the peace of this landscape at least, the personally conducted tripper will not intrude just yet.

I noticed several cataracts, but none of any great extent. The water in these rapids was rushing over and between rocks, and they must be ugly places when the lake rises. In other parts the river was smooth and still. In one pool we counted eighteen hippos taking the air, just their snouts being out of water.

Almost at once we set to work and piled up a heap of stones, and fixed in the centre of it a long pole as a landmark. This was a straight, thin branch which we cut from a tree near at hand. After this we set up three more "cairns" on selected spots, and then descended to the low ground, where the mules were waiting. I remained in the valley, for exertion soon tires one out after the fever. My friends climbed other hills and erected more landmarks before returning to lunch. In the meanwhile I tried my luck at angling in the river, but caught nothing, and soon gave up the attempt, and lounged under a palm-tree, reading.

My companions started on their mules after lunch to continue their survey, and I went back to camp and was glad to be idle. The day's work was ended before evening, and I believe we were the first Britishers who had five-o'clock tea beside the head waters of the Blue Nile.

Next morning (January 27) I had instructions to set up a stone-heap on a little hill about four miles from camp. I made my way to it direct through a tangle of jungle-growths, and from the summit obtained the photograph of the source of the river at the point of outflow from the lake which forms the frontispiece of this book. I believe this to be the only photo of the place in existence.

I descended the hill, and walked along the river-bank towards the place where we had arranged to lunch. The soil was boggy, covered in some parts by reed-grass and in others by jungle-growths. Hippo spoor was everywhere. This tract is not well-wooded; there are a few palms, and some scattered trees of other kinds, chiefly mimosas. The bed of the river is hard rock, and the water was clear and drinkable. The depth varied very greatly, from a few inches in the rapids to some six metres in the pools. I saw the stream at its lowest. Watermarks showed clearly that in flood time it rises sixteen to eighteen feet above this level.

When I reached the place where we were to lunch, I saw a big herd of hippos basking on the surface of the river. Crawley and I walked towards them, and when they saw us, they sank, leaving only their snouts visible. These offered an interesting target for rifle-fire, and for awhile the hippos watched our practice with unconcern. Then a shot told—there was a "general post"—and not a sign of the huge beasts remained in sight.

After this I took my rod and tried for a fish in a pool below a rapid. While I was standing there, I saw a large white eagle, a splendid fellow, which had been watching me fishing, swoop from the branch of a high tree. He circled twice above me, and then pounced upon a young duck, that was hiding under a ledge in the river-bed, and bore it off. The parent birds were close at hand, and I heard their loud, frightened, and indignant cackling. I do not think wild duck are plentiful hereabouts. No doubt they are harried by these eagles and have to find safer breeding-places.

We saw only a very few. I landed a plump fish of the perch tribe, which weighed about four pounds. He showed no fight, but came tamely into shallow water, where the boy who was with me picked him up. My friends came late into camp after a hard day's work upon their survey.

The high ground beside the head waters of the Blue Nile is pleasant and healthy and could support a large population. The maximum temperature in the day, when we were there, was about eighty-five degrees. Usually a cool, exhilarating breeze was blowing, and we did not feel the heat disagreeably. If this place were more easily accessible, it would be a perfect pleasure-resort and a most valuable sanatorium for residents in the Soudan.

The event of January 28, was the appearance of potatoes at table. I had often gazed at this unvarying factor in one's diet with indifference if not slightingly, and had wondered why the things were served so constantly at European tables. It was not till we missed the daily luxury that we appreciated it. The roast fowl was another bird with this accompaniment. These were the only potatoes that we saw in Abyssinia; they were very small. The cook had discovered them in a remote hamlet. I can offer no explanation of their presence in that place. He purchased a sackful for one salt and an empty lime-juice bottle, and returned to camp in triumph, shouting at the top of his voice that he had found "batatas" at last. Perhaps as he failed to discover any more, the expedition denuded the country of its whole stock.

CHAPTER X

On January 29 my companions and I started early for the little hill beside the river on which I had set up a "cairn" the day before. This is the first rising ground beyond the outlet of the Blue Nile. (It ought to be named after some kingly, heroic, half-divine figure that looms in the dawn of legend, but—with my modest compliments to future geographers—it has since been called Hayes's Hill.)

At the foot of it, just below the first cataract in the stream, is a ferry. It is a primordial means of transit into Godjam. Two men are in charge of a papyrus boat which they paddle across, using a pole in the manner of a canoe-paddle, as the water is too deep for punting. At this spot Dupuis had some survey work in hand which involved taking a rope to the opposite bank. Our Berthon boat had been brought down to the riverside, and Crawley and I rowed over towing the line astern and made it fast to a tree. Crawley returned, and I remained at leisure in Godjam and used my camera and fishing-rod to wile away the time.

I saw some natives using a net like a seine in the shallow water just below the rapids. They had some very fair catches, and among the fishes which they hauled ashore I noticed in particular one of a blue colour which had what appeared to me to be a red sucker on its snout. I have no idea to what species it belonged.

My friends finished their mapping about one o'clock, and then we lunched in the shade of a mimosa. While we sat there we had an opportunity of watching the Habashes' method of getting donkeys across the stream. They were pushed into the water, then four to six men would scramble on to the crazy *tankoa,* which looked as if it must sink under their weight. Each man would now seize a donkey by the ear, and then the boat would start guided by the old ferryman in the stern. Thus the donkeys had to swim alongside the raft—it is no more than that—and as their heads were held above water, they could scarcely come to harm.

We saw that it would save us much time and trouble if our beasts were towed across in this way, and, after some deliberation, we asked the ferryman what his charge would be for transporting seventy donkeys. He replied that he would do it for seven dollars. This, on consideration, seemed a fair bargain, and we resolved to make an early start on the following day. The baggage was to go over in our own boat, which could be hauled across by means of a line made fast on both banks.

After concluding this arrangement we went back to camp, carrying the rope with us. It might have tempted Charon—I don't know his Abyssinian name—if we had left it. We had sent Walda Mariam as an envoy to King Tecla Haimanot, whose lands we should enter after crossing the river. The king was in residence not far off, and in the evening our man returned with a satisfactory message giving us all necessary permission. The jurisdiction of Tecla Haimanot extended only ten or twelve miles northward beside the lake, and we found

THE DONKEYS CROSSING THE FERRY. [*See p.* 140.

THE MULES SWIMMING THE FERRY. [*See p.* 140.

that we should have to obtain the consent of Ras Mangousha, whose dominions march here with those of the King of Godjam, before we could complete the circuit of the lake.

We made an early start on the morning of the 30th. I went to the ferry and had our Berthon boat put together. We found that our plan answered excellently. Our boat, loaded with baggage, was easily hauled over, and by a quarter to ten half a score of donkeys had been carried across in tow of the ferrymen's *tankoa*. Some of the boys would put their arms under a donkey and lift it bodily into place by the ferry-boat at the starting-point, where the water was shallow. It was a scrambling, pushing, splashing business, and they thoroughly enjoyed standing in the stream and basking in the sunshine. A number of Habashes from Woreb helped, and our crossing was a merry, pleasant scene. The mules did not go "by ferry," but swam over with men beside them, as at other deep water passages. It was all done and we had landed in Godjam without mishap of any kind by a quarter-past one. The last load that I took over in our boat consisted of nine of our men, rather more passengers, I fancy, than I was "licensed to carry." We scrambled up some rocks on the further side, and then found ourselves on level ground, where travelling would be easy.

Fully half of the donkeys had been reloaded by the time we finished lunch, and we moved ahead with this detachment. The country on the further side of the Nile seemed to me to be in a more prosperous state. There was more cultivation of dhurra and grain. The natives were fatter and looked, by comparison, "well-to-do." I heard afterwards that the ruler here is not so extortionate as certain other chieftains. I believe that we were the first Europeans to traverse this part of Godjam.

After a journey of three-quarters of an hour we approached the village of Bahardar Georgis. Before entering the hamlet we had to conform to a singular usage. Two men had stationed themselves beside the track, one on each side of it, and they held a *shama* across it. We asked Johannes what this meant, and he told us that it was to protect the villagers from the power of the "evil eye." This is lost if the stranger who may possess it passes under the *shama.* We had to move the greasy robe aside and go beneath it, hoping rather than believing that it was not verminous. The Habashes are extremely superstitious in this respect. It is customary to screen a person of rank with a shama when he drinks, to safeguard him from the same peril, and such persons are frequently kept from view likewise while they eat their meals. Mr. Hormuzd Rassam, being a high dignitary, was carefully shielded in this way to his no slight disgust. He wrote: "As the generality of the garments spread out for our seclusion had not been washed for months, and probably not since they were first worn, the reader's imagination may be left to conceive the odour which surrounded us on these occasions. But even if they had been washed no later than the previous day, the disagreeable smell of rancid butter with which the natives besmear their heads would suffice to render any such curtain almost intolerable." *

We had to use care in the selection of our camping-ground near Bahardar Georgis. Much of the ground is covered by the papyrus plant, and this shows

* "Narrative of the British Mission to King Theodore," vol. i. p. 229.

PUSHING THE DONKEYS INTO THE WATER PREPARATORY TO THEIR BEING FERRIED ACROSS. [*See p.* 141.

dampness and, by consequence, risk of fever. We finally selected a spot on some high ground, where there was a dry, red soil. This happened to be close to a little settlement of the curious people called *Waitos*, who are only found in this district. They inhabited two or three huts near the village. Mr. Rassam has given a concise and interesting account of them: "The Waitos are Mussulmans of the Maliky sect, and although Mohammedanism recognizes no castes among its adherents, nevertheless these people, who subsist upon the flesh of the hippopotamus" (which is considered unclean by all other Abyssinians), "are looked down upon by their co-religionists, who consider it a degradation to associate with them. A few among them cultivate a little grain, but the flesh of the hippopotamus forms their staple food. . . . I was unable to obtain any satisfactory account of the origin of this peculiar people. It is just possible, however, that there may be some relationship between them and the *Watos*, a tribe of Gallas inhabiting the banks of the Hawash, south of Shoa, who are also said to live on the flesh of the hippopotamus." * Stecker, who was scrupulously accurate in almost every particular which he mentioned, said that the *Waitos* were, "strictly speaking, a Pagan sect (*eigentlich Heidensecte*)," † but in this he was mistaken. Oddly enough, these people, though they are, in a sense, outcasts, are exclusive, and proud of their isolation.

On the morning of January 31 my companions took the Berthon boat and a whole cargo of surveyors' gear, and started to complete the work for which

* "Narrative of the British Mission to King Theodore," vol. i. p. 314.

† "Mittheilungen der Afrikanischen Gesellschaft in Deutschland," 1881, vol. iii. p. 27.

the expedition had been organized. A full account of the results will be found in Sir William Garstin's official Report on the Sources of the Nile. My services were not requested, and I certainly have had more practice in surveying symptoms than ground. I stayed in camp for awhile, and then walked to the river with angling-tackle, and landed a fair-sized fish with a spoon-bait. It was of the perch family.

The country around Bahardar Georgis is flat, as I have said, and in places swampy. The marshy tracts are overgrown by papyrus, and it is not easy to distinguish them before one is both on and in them. On this moist ground we saw flocks of wild geese and various species of ibis. In the neighbourhood of the village I first observed the bird known as *buphaga Africana.* It pitched upon the backs of any of our donkeys that had sores, and caught the flies that settled there. The bodies of these birds are gray, and scarcely distinguishable from the donkey's hide. Their beaks are red, and, at a little distance, I noticed that they exactly resembled a sore on the beasts which they infested. This colouring seems a remarkable example of evolution in the direction of "protective mimicry." They worry cattle also, and it is well known that they peck the animals to make raw places, which then attract the flies that are caught upon them. I believe that they also use their beaks to mimic sores, for my attention was first drawn to the birds by the semblance of abrasions on the spinal area of some of the donkeys, where I knew that none existed. No doubt the beaks, used thus, are a successful trap.

During the day I received a visit from an Abyssinian artist who lived at Bahardar Georgis. He came to beg colours from us, if we had any; I have no notion how the Habashes compound their pigments. He told me that he had a commission to execute some paintings for the church. I could only give him some pieces of red and blue pencil. He was extremely grateful for these, and bestowed upon me some samples of his art.

I presented him to Dupuis, who was greatly interested by his pictures, and kindly took the trouble to discover, after much rummaging, an old box of paints. This he gave to the youth, whose delight was indescribable. He promised Dupuis works of art in his best style. In the evening he came to our camp again, and I showed him the Christmas Number of *Pearson's Magazine.* This pleased him highly, especially when he found that he was to be the possessor of it. The three coloured plates caused him an ecstasy of wonder and pleasure, and the respective artists—and the colour-printers—have a venerating admirer by the lake-side.

My friends returned to camp rather later than they had expected, but were well pleased by the result of the day's work. They found that at this season of the year the discharge from the lake into the river was at the rate of forty-two cubic metres to the second. And they brought some guinea-fowl to the larder. The future proprietor of the Blue Nile Hotel and pension on Lake Tsana will always be able to offer his guests fish and poultry.

On February 1 my companions started on foot, accompanied by Johannes, a pack-mule and some boys, to see the first falls in the course of the Blue Nile, and the ancient stone bridge which spans the river at this point.* These falls are twenty-one miles below the outlet of the stream from the lake.

* See p. 151.

I remained in charge of the camp, and, to wile away the time, took our boat out upon the lake. This caused a great sensation among the villagers, who had seen nothing of my friends' excursion in it the previous day. The Habashes flocked to the edge of the water, but whenever I rested on the oars for a few moments, they rushed screaming into their huts. Apparently they regarded me as a naval force, and thought I meant to carry Bahardar Georgis by a *coup de main*. Perhaps they remembered King Theodore's descent upon the island of Dek. "He was in pursuit, it appears, of a refractory chief under Ras Ali, who had taken refuge on the island. In less than twenty-four hours he had two hundred canoes constructed, in which he suddenly appeared off the place with five hundred chosen warriors." * I should think that this was the only occasion on which the *tankoa* was used in war, and have a feeling of compassion for the five hundred warriors.

We noticed that at this place the children seemed especially frightened of white men. Perhaps their mothers used us as bogeys to terrify those that were troublesome. We found that we could soon regain the confidence of the people by giving them any little picture, no matter of what; prints from advertisements of furniture or clothes served quite well. They received these with delight.

On the next day (February 2) my friends returned to camp, having made short work of their tramp of forty-two miles. Crawley had shot an oribi, a species of gazelle, on the way to the falls, and Dupuis secured one on the march back. The former had served as tasty

* Hormuzd Rassam, "British Mission to King Theodore," vol. i. p. 315.

rations to the party on the journey, the latter came into our larder, and the prospect of venison after a long course of guinea-fowl was very pleasant.

They had met a small body of Habashes on the road, who attempted to turn them back, and these men loaded their rifles as a menace. They were probably soldiers of the King of Godjam, but as the Abyssinian troops wear no uniform it is difficult to distinguish those who are "in the service" from those who are not. As among the Boers, the combatant can at any time become a non-combatant if he has time to hide his rifle and cartridges. My friends took no notice of the hostile demonstration and rode on. Nothing came of it. Probably the threat was intended to extort money. Usually we were well received, and I attribute this to the sound judgment of our chief, who took care to leave the villages near which we camped a little richer than we found them.

The bridge is a most interesting relic of the times of Portuguese ascendency in Abyssinia. I am indebted to Dupuis for permission to reproduce his photograph of it. The following account of its construction is taken from Dr. Johnson's translation of Father Jerome Lobo's * "Voyage to Abyssinia." I quote the passage *in extenso* because it contains a reference to a question which has been much discussed, though it has merely an academic interest, viz. whether Lake Tsana, or the river which is its principal tributary, should be regarded as the true source of the Blue Nile.

"The Nile, which the natives call Abavi" (Abai), "that is the father of waters, rises first in Sacala, a province of the kingdom of Goiama" (Godjam), "which is one of the most fruitful and agreeable of all the

* Father Lobo reached Abyssinia in 1624.

Abyssinian dominions. This province is inhabited by a nation of the Agaus, who call themselves Christians, but by daily intermarriages they have allied themselves to the Pagan Agaus, and adopted all their customs and ceremonies. These two nations are very numerous, fierce, and unconquerable, inhabiting a country full of mountains, which are covered with woods, and hollowed by nature into vast caverns, many of which are capable of containing several numerous families and hundreds of cows. To these recesses the Agaus betake themselves when they are driven out of the plain, where it is almost impossible to find them and certain ruin to pursue them. This people increases extremely, every man being allowed so many wives as he hath hundreds of cows; and it is seldom that the hundreds are required to be complete.

"In the eastern part of this kingdom, on the declivity of a mountain, whose descent is so easy that it seems a beautiful plain, is that source of the Nile which has been sought after at so much expense of labour, and about which such variety of conjectures hath been formed without success. This spring, or rather these two springs, are two holes, each about two feet diameter, a stone's cast distant from each other. The one is but about five feet and a half in depth, at least we could not get our plummet farther, perhaps because it was stopped by roots, for the whole place is full of trees. Of the other, which is somewhat less, with a line of ten feet we could find no bottom, and were assured by the inhabitants that none ever had been found. It is believed here that these springs are the vent of a great subterraneous lake; and they have this circumstance to favour their opinion, that the ground is always moist, and so soft that the water boils up underfoot as one walks upon it:

this is more visible after rains, for then the ground yields and sinks so much, that I believe it is chiefly supported by the roots of trees that are interwoven one with another. Such is the ground round about these fountains. At a little distance to the south is a village named Guix, through which the way lies to the top of the mountain, from whence the traveller discovers a vast extent of land, which appears like a deep valley, though the mountain rises so imperceptibly that those who go up or down it are scarce sensible of any declivity.

"On the top of this mountain is a little hill, which the idolatrous Agaus have in great veneration. Their priest calls them together at this place once a year; and having sacrificed a cow, throws the head into one of the springs of the Nile; after which ceremony every one sacrifices a cow or more according to their different degrees of wealth or devotion. The bones of these cows have already formed two mountains of considerable height, which afford a sufficient proof that these nations have always paid their adorations to this famous river. They eat these sacrifices with great devotion, as flesh consecrated to their deity. Then the priest anoints himself with the grease and tallow of the cows, and sits down on a heap of straw on the top and in the middle of a pile which is prepared. They set fire to it, and the whole heap is consumed without any injury to the priest; who, while the fire continues, harangues the standers by, and confirms them in their present ignorance and superstition. When the pile is burnt, and the discourse at an end, every one makes a large present to the priest, which is the grand design of this religious mockery.

"To return to the course of the Nile. Its waters, after the first rise, run to the eastward for about a musket-shot; then turning to the north, continue hidden in the grass and weeds for about a quarter of a league, and discover themselves for the first time among some rocks; a sight not to be enjoyed without some pleasure by those who have read the fabulous accounts of this stream delivered by the ancients, and the vain conjectures and reasonings which have been formed upon its original, the nature of its water, its cataracts, and its inundations, all which we are now entirely acquainted with, and eye witnesses of."*

The Nile "rolls away from its source with so inconsiderable a current that it appears unlikely to escape being dried up by the hot season, but soon receiving an increase from the Gemma, the Kelta, the Bransu, and other less rivers, it is of such a breadth in the plain of Boad, which is not above three days' journey from its source, that a ball shot from a musket will scarce fly from one bank to the other. Here it begins to run northward, deflecting, however, a little towards the east, for the space of nine or ten leagues; and then enters the so much talked of lake of Dambia" (Tsana), "called by the natives Barhar Sena, the resemblance of the sea, or Barhar Dambia, the sea of Dambia. It crosses this lake only at one end, with so violent a rapidity that the waves of the Nile" (Abai) "may be distinguished through all the passage, which is six leagues.† Here begins the greatness of the Nile.

* Father Lobo, who landed on the east coast of Africa, had never heard of the White Nile—this is made evident by his discourse upon the Abai—and believed the head-waters of the Blue Nile to be the origin of the whole river.

† In the map reproduced in this volume, the course of the Abai through Lake Tsana is marked by arrows. Its current can be quite plainly distinguished, as Father Lobo says, from the surrounding water. The circumstance was remarked by Dr. Stecker.

Fifteen miles farther, in the land of Alata, it rushes precipitately from the top of a high rock, and forms one of the most beautiful waterfalls in the world. I passed under it without being wet, and resting myself there for the sake of the coolness, was charmed with a thousand delightful rainbows which the sunbeams painted on the water in all their shining and lively colours. The fall of this mighty stream, from so great a height, makes a noise that may be heard to a considerable distance. . . . The mist that rises from this fall of water may be seen much farther than the noise can be heard. After this cataract, the Nile again collects its scattered stream among the rocks which seem to be disjoined in this place only to afford it a passage. They are so near each other, that, in my time, a bridge of beams, on which the whole Imperial army passed, was laid over them. Sultan Segued * hath since built here a bridge of one arch in the same place, for which purpose he procured masons from India" (*i.e.* the Portuguese Indies).† "This bridge, which is the first the Abyssinians have seen on the Nile, very much facilitates a communication between the provinces, and encourages commerce among the inhabitants of his Empire."

Bruce's description of the source of the Abai differs somewhat from that given by Father Lobo. It is brief,

* See p. 256.

† Stern mentions "a solid stone bridge of seven arches—one of the few monuments left by the Portuguese," which in his time spanned the Reb. He called the stream Erib ("Wanderings among the Falashas," p. 169). Both he and Dufton crossed that river by this bridge. In all probability it was constructed by the artificers who built the arch over the Abai. I do not know whether the Reb bridge is still standing. We forded the stream, as has been said, and were not aware at the time that it had ever been bridged.

and as an exceptional interest attaches to the spot, it may not be superfluous to quote it.

"Half undressed as I was, by loss of my sash, and throwing my shoes off" (as he had been directed to do in order to conform to a superstitious usage of the natives), "I ran down the hill towards the little island of green sods, which was about two hundred yards distant; the whole side of the hill was thick grown with flowers, the large bulbous roots* of which appearing above the surface of the ground, and their skins coming off upon treading upon them, occasioned me two very severe falls before I reached the brink of the marsh. I after this came to the altar of green turf, which was in form of an altar apparently the work of art, and I stood in rapture over the principal fountain, which rises in the middle of it." †

Bruce talked with "the Shum, the priest of the river, whose title was Kefla Abai, or 'Servant of the river.' He was a man of about seventy. The honourable charge which he possessed had been in his family, he conceived, from the beginning of the world; and as he was the happy father of eighty-four children, it appeared that his race was likely to flow as long as the Nile itself. He had a long white beard; round his body was wrapped a skin, which was fastened by a broad belt. Over this he wore a cloak, the hood of which covered his head, his legs were bare, but he wore sandals, which he threw off as soon as he approached the bog from which the Nile" (Abai) "rises—a mark of respect which Bruce and his attendants were also

* The bulbous roots here mentioned are probably those of the "red-hot poker" plant, which grows abundantly in damp places in the higher regions of Abyssinia.

† "Travels to discover the Sources of the Nile in the years 1768, 1769, 1770, 1771, 1772, and 1773," by James Bruce of Kinnaird, Esq., F.R.S.

required to perform." The Shum very obligingly presented his comely and sprightly daughter Irepone, aged sixteen, to the Scottish traveller as housekeeper.

Bruce says that at this time the people of the place called the Spirit of the River "The Everlasting God, Light of the World, Eye of the World, God of Peace, Saviour and Father of the Universe." He asked the Shum if he had ever seen the Spirit, and the old man answered without hesitation, "Yes, very frequently." *

The Abai mentioned above is the main tributary of the Blue Nile, but a glance at the map will show that Lake Tsana receives many other rivers, and the surrounding mountains, of course, add the volume of innumerable torrents and small streams to its waters in the rainy season. For practical purposes the source of the Blue Nile is the lake fed by all the affluents which collectively determine the amount of the discharge into the river.

To return to the camp at Bahardar Georgis. The survey work of the expedition was now completed, and our subsequent stages were upon the homeward journey. On February 3 we visited the village church. I obtained a clear photograph of some of the paintings in the interior. Unfortunately, as I am no archæologist, I cannot pretend to say whether the clothing of St. George, either in his combat or his victorious return, or the conception of the other figures, gives an indication of the origin of the Abyssinian school of ecclesiastical art. I hope that some of my readers who are better informed may be able to throw light upon the subject. It is, perhaps, of considerable interest for the following reason: If the usages of the Abyssinian church, which

* "Life of Bruce," by Major F. P. Head, pp. 276–278.

is strictly conservative, represent a really primitive form of Christianity, they show that the observance of ritual ceremonies was, in the early days, at least as much a matter of concern as the condition of the individual conscience. And the style of the Abyssinian pictures of sacred subjects may help to determine the date when the accepted liturgy took its present form.

In the afternoon the guns of the party added some venison, poultry, and game to the larder, and in the evening we were serenaded—against our will and at our expense. A band of singers and dancers from the village—both men and women—came into camp. They had no intention of showing their skill without remuneration, and as it would have caused ill-will among them and disappointment to our own boys and escort if we had sent them away, we endured their performance and paid for it. One of the instruments which they use is a piece of board, over which strings are stretched, so that it looks like a rude archaic forerunner of the violin. They twang the strings with their fingers, but do not "stop" them to obtain different notes from the same string. Our troupe also played the tomtom and sang in the high nasal voice which is characteristic of the race. We heard the last of them, thankfully, at half-past nine.

While we were at Bahardar Georgis, the Waitos near our camp drove a brisk trade in courbashes.* I bought of them some hippo tusks, which they were glad to sell, as they live in great poverty. If I had had some more small change (salt) I could have purchased a quite considerable stock.

* Whips made of thongs of hippopotamus hide, which serves both for lash and handle.

INTERIOR OF THE CHURCH AT BAHARDAR GEORGIS. [*See p.* 153.

Insects are an annoying pest in this part of the lake side. Hosts of mosquitoes and small beetles of a peculiar species appeared on the flat ground at sunset, and swarmed throughout the neighbourhood for an hour afterwards. The beetles settled upon us in throngs, and crawled into our noses and ears and under our clothes, and we could not even crush them on account of the unendurable smell which their bodies then gave forth. When the first hour after sundown is past, the temperature falls very rapidly, and one sees no more of these insects.

On February 4 we started to complete the circuit of the lake and reach Delgi by way of the Zegi peninsula. Our road lay across flat land, bordered by marshes and full of swamps and quagmires. Sometimes the marshes stretch a long way into the land, and long detours had to be made to avoid them. I saw yams for the first time in Abyssinia growing just above the swampy tracts in this region. We had to cross three rivers with rocky beds, which were made rougher and more slippery by loose stones. To add to our difficulties, our guide twice led us out of the right path, and once to a ford which was impassable for the donkeys. In this way we lost nearly three hours. At lunch time we sent the baggage-train on ahead of us. The latter part of the journey was over a beaten track, and gave us no trouble. We overtook the baggage animals and their escort just as they entered the undulating ground which forms the approach to the peninsula of Zegi. We pitched our camp near the shore of a little bay of which this promontory was the further boundary.

While we were on the road I received a scribbled note from Crawley, who told me that one of his soldiers was ill, had lain down and refused to move.

I rode back at his request, and found the invalid under a tree. He said, "Leave me alone. I want to die." It was evident at a glance that he was suffering from ague. The only remedy which we had at hand was chartreuse. I gave him a big "nip" of the cordial, and it had an excellent effect upon him. He was able to ride to the end of the journey, and was none the worse for the effort. I venture to commend this incident to the consideration of strict teetotalers.

The village near which we were encamped is that which is marked as Furje on Stecker's map. The district affords a curious example of feudal tenure in Abyssinia. We had quitted Tecla Haimanot's dominions, and the land on which our camp stood was under the control of a certain chief called Fituari * Ali, a feudatory of Ras Mangousha. He dwelt close to the town of Zegi, but had no jurisdiction within its boundary, though his lordship was valid in a region extending beyond the town to the Abai.

The chieftain had gone to attend the marriage of Ras Mangousha's daughter, and had left his son in authority. So we sent a messenger with an escort to carry the news of our arrival to this young Habash with due formality. He brought back an uncivil reply to the effect that the Fituari's son was absent, and if we wanted anything we had better go and find him. This was sent by his majordomo. While we were waiting for tea to be served, Johannes reported that the young Habash was approaching, and we saw him at a little distance attended by a band of followers, some of whom carried guns. Our interpreter asked what he should say to this truculent young man, and we bade him

* This is a military title, and means "Commander of Advance Guard." (A. Wylde, "Modern Abyssinia," p. 495).

explain that we only asked leave to pass through the land, and should require nothing unless it were to purchase a little grain for our animals. We always sought to avoid trouble with the natives, and therefore impressed upon Johannes that he should show we wished to be friendly, and say we hoped the Fituari's son would come and have a drink with us.

Johannes departed with his message, and presently we heard a great hubbub—many Habashes talking at once at the top of their high-pitched voices. We wondered what gave rise to so much excitement. Presently Johannes emerged from the crowd and approached us slowly. The young man's answer was that he would speak with us when he had seen the King's letter. Now, this permit and all our credentials had been dispatched on February 1 from Bahardar Georgis to Ras Mangousha that we might obtain his leave to travel through his territories beyond the Abai, and we did not expect our messenger to return until late on the following day.

It was an uncomfortable situation. The Ras's reply might be unfavourable. In that case we should be confronted by the necessity of retracing our steps over the whole of the toilsome journey by the lake side. We all longed to kick the tiresome coxcomb who was in our way, and went to dinner in a glum mood.

We were obliged to spend the following day (Feb. 5) in inaction awaiting the return of our messenger with Ras Mangousha's answer. I busied myself with the camera, having every reason to believe that no photographs of this tract of country had ever been taken.

During the morning we received a visit from the head man of Zegi. We thought it a favourable sign that he gave us a very pleasant and courteous welcome. This young man, Hyli by name, was about nineteen years of age. I learned afterwards that he was studying the ancient Geez language under the tuition of the priests of Zegi, and presume that he intended to "go into the Church." These candidates for orders are not permitted to smoke or drink strong liquor while they are *in statu pupillari*. A similar restriction would scarcely be popular in our own ancient universities.

Hyli, we found, had a large consignment of coffee to send to the market at Gallabat. It is his business to collect the dues payable on this produce before it leaves the village, and the revenue so obtained is handed to Ras Mangousha. Hyli had now come to request that his caravan might join ours during the journey through the "rain-country"—that borderland between Abyssinia and the Soudan, which, as I have said, is infested by bandits. We had every reason to win friends where we could, and every wish to please the young Habash, so we consented willingly. He told us that the coffee was already at Delgi, and that he had been informed of the date of our arrival at that village, at Korata and at Woreb, and had been looking forward to our coming for a month past. In the evening he sent us a present of flour and fowls.

After this visit, I walked to the township of Zegi. It is surrounded by a thick hedge of incense-bush, and this forms the boundary between the Fituari's jurisdiction and Hyli's. Zegi very closely resembles Korata. It consists of groups of tokhuls scattered among small, square enclosures where the coffee bushes grow apparently untended. These plantations, with the cottages and churches among them, cover the whole promontory. I should estimate the population, when I saw the place, at about three thousand souls.

Dr. Stecker's account of his visit to the town is brief

and interesting, and I quote it. I saw nothing of the stone dwellings which he describes, and think they must have been replaced by straw tokhuls since 1881. He wrote:—

"On June 7 I made a *tankoa*-journey to the peninsula of Zegi, and climbed to the highest peak, Tekla Haimanot (2074 metres above sea-level, according to barometrical measurement), which afforded extremely important survey-bearings." The traveller then mentions his visit to Livlivo, Adina, and the island of Dek, and adds, "The Zegi peninsula is especially famed for its coffee plantations. Some coffee-trees are as much as a metre in girth. The coffee is mostly exported to Metemmeh" (Gallabat), "less goes to Massowah, but it is not considered so good as that of Korata. Besides coffee the *Ensete* banana flourishes here conspicuously, and also the edible species (*Musa Ensete edulis*); but, unfortunately, in recent years these charming plantations have been almost entirely destroyed by a species of pig called *Assama* (*potamochoerus penicillatus*),* which is found here in hundreds. This remarkable animal feeds almost entirely on the roots of these fine bananas. What struck me here particularly was the neatness of the tokhuls, which are chiefly of stone, and in general all villages on Lake Tsana have a much cleanlier and more pleasant appearance than those inland. There is no lack of clergy on the Zegi Peninsula: there are here no fewer than seven churches with twelve hundred priests and defterers." † I am bound to say that I saw no indication during my brief stay that the population was deplorably priest-ridden!

* We saw few bananas and none of these pigs in 1903.

† "Mittheilungen der Afrikanischen Gesellschaft in Deutschland," 1881, vol. iii. pp. 30, 31.

I was returning to camp about four o'clock in the afternoon, and was still at some distance from it when I met a Habash, who made me understand by signs that our messenger had brought the Ras's letter, and as I hastened on I noticed that the news was already public property. Upon reaching camp I saw our man, grimy and travel-stained. He and a companion, with one mule to ride, had covered about a hundred and thirty miles in four days over very rough country, and they had waited while the Ras attended to our business; so they had not let the grass grow under their feet. I felt sorry for the mule. Walda Mariam had had charge of this business. We had given him one day's rest at Bahardar Georgis after his return with Tecla Haimanot's message, and then dispatched him on this second journey. It is expected of these runners, when they are in charge of a missive from the Negus or a great chief, that they shall not sleep till they have delivered it. The man bowed low, and handed the Ras's letters to me in a manner which showed that he now made me responsible for their custody. I then learned, by the aid of an Arab interpreter, that the chief's reply was of the most favourable kind, and that he had sent mandates to all concerned to give us every furtherance on our way round the lake. He also inquired very courteously about our health and our progress, and had sent a soldier from his own guard as a special escort for the party. The Ras, moreover, had even furnished us with letters to chiefs through whose lands we should not pass on the road to Delgi, to be used in case we wished to turn aside from the way and visit the *hinterland* of the lake district. And, best of all, there was a communication addressed to Fituari Ali's son, enjoining upon him that he should show us every civility. Johannes, who had been absent from camp when the messenger arrived, had returned by the time my companions came back from an excursion. The despatches were then interpreted to them in French, and we enjoyed the prospect of our enemy's discomfiture. It was resolved that the mandate to him should be delivered on the following morning. I noticed that the Habashes did not appear to make common cause with Ali's son, but seemed pleased at our success. Among the Ras's letters was one to Hyli, which we sent to him immediately, though it was scarcely required in his case. Zody was the bearer of it.

CHAPTER XI

February the 6th was a market-day in Zegi. In the morning we mounted our mules and went to visit St. George's Church. Hyli was studying in the theological school attached to this round, thatched place of worship, which resembled in all respects the others that we had seen in the country. When the Chief Priest had received a suitable offering, our Abyssinian friend took us to his house, and here, for courtesy's sake, we were obliged to drink *tedj*—a vile, bitter draught. We had escaped it on other occasions. The composition of it has been mentioned elsewhere. Hyli was very anxious that we should stay and eat a meal with him—doubtless it would have consisted chiefly of raw meat smothered with red pepper and sour *teff* bread—but we managed to excuse ourselves from this ordeal. On our way back to camp we passed through the market-place, which was now thronged. I do not know whether any European had been seen there before; but in any case we were objects of the utmost curiosity, and the people pressed around us so thickly that we had a difficulty in making our way through them.

In the afternoon I returned to the market for the purpose of obtaining snap-shots. It is held on the top of a stretch of rising ground, under the shadow of some half a dozen big trees. Under each tree was a large

MARKET DAY AT ZEGI. [*See p.* 162.

stone. On market-days a priest from each of the different churches stands on the stone allotted to his parish. These men are striking figures, clad in their ecclesiastical vesture, of which a large white turban and a shama with a broad red border are the conspicuous features. The parishioners from the different districts squat around their pastor near the stone, and the priest takes tithe in kind upon the spot when any member of his flock completes a purchase or a sale. Our people bought some grain and two sheep. The ruling prices were: for an ox (without the hide), seven shillings; for a sheep, two shillings.

In one of the photographs which I obtained, the curious plaits in which the married women wear their hair are clearly shown. Stern thus describes the manner in which the *coiffure* is preserved from disarrangement at night. "The woman whose hair has undergone the tedious process of plaiting, must also have it protected from becoming dishevelled while she sleeps; and as this cannot so easily be done in a country where a bullock's hide or a mat forms the bed, necessity has contrived a bowl-shaped stool, in which the neck is wedged. . . . In Abyssinia, where the women are particularly proud of their copper-coloured charms, very few, even on a journey and with fifty pounds weight on their backs, will forget to take the wooden pillow and the hollow grease-filled gourd," from the contents of which the hair is "dressed." *

I purchased a leopard's skin in the market for a dollar, but it was not a good specimen. During the afternoon I received a visit from another Abyssinian artist, who presented me with two pictures in return for a black lead-pencil and a part of a blue one. And I

* "Wanderings among the Falashas," p. 13.

had a constant stream of patients, who claimed attention very freely. I am bound to say that the maladies from which the majority of them suffered fully justified the allegation of Dr. Stecker, to which I have already referred.

The letter from the Ras to the Fituari's son was delivered early in the morning. In the evening we heard that no sooner had he received it than he disappeared, and nobody seemed to know his whereabouts. No doubt it was his intention to declare afterwards that he had not been in the village when we arrived. If his conduct came to the Ras's knowledge, I have little doubt the vainglorious youth was flogged—this has been the penalty inflicted on other Habashes who have shown rudeness to travellers provided with the King's safe-conduct letter.

Our tents at Zegi were in a pleasant position, under a spreading fig tree; these trees are found throughout Abyssinia and in the "rain country," and give abundant shade. I have never seen them growing thickly, in a clump.

We made an early start on the morning of February the 7th, and trotted ahead at a good speed, as the donkeys were very fresh after their rest. The country is similar to that which we traversed in approaching Zegi. At eleven o'clock we reached the bank of the Abai. At the ford where we were to cross, it is a broad river, more than a hundred yards in width, as I should judge. The water was running in a fairly rapid current, and I was told that the stream is perennial. The banks of the river are steep, and the bed is stony. The water, at that season of the year, was almost clear. But during the rains, when the stream is in flood, it brings down vast quantities of the deposits of the white ants and other detritus. The flat island of Dek has been formed by siltage of alluvial soil thus brought, and it is matter in solution which renders the course of the Abai traceable in Lake Tsana.

At the ford we found that the water came just above a man's knee. The crossing gave us little trouble, and there was no serious mishap. The larger loads and our valises—of the "Wolseley" pattern—were wetted, but the sun soon dried them. One donkey collapsed, and fell with his burden into the stream about two yards from the further bank; but there were many to help, and he was soon put on his feet again. He was not carrying anything which would be ruined by a soaking. The rise beyond the ford was steep, and the drippings from the wet animals made it slippery, so we had to throw earth on it to give them a foothold, as in crossing the Gelda.

After passing the Abai, we entered a flat district full of the long dry grass, of which we had seen so much on the north side of the lake. The country hereabouts is full of the kind of gazelle called *oribi*. We had not elsewhere in Abyssinia found these creatures in herds. Game birds abound in this region, which appeared to be almost deserted. We passed scarcely any villages, and those which we saw consisted of only five or six huts.

Our camping-ground was an open space not far from a papyrus swamp. We should not have selected this spot by choice, as the proximity of marshy soil was a danger to health. But we were obliged to halt there because no water was to be had for a considerable distance on the road ahead. All around were ruins of houses built of stone, with thatched roofs that had fallen in. The number of them showed that a town or a large village must have existed here at one time. Stone dwellings are not usual in this region, and I inquired what the name of the place was and tried to learn its history, but could get no information.

Dupuis and Crawley went out with their rifles, and added three oribis to the store in the larder. I stayed in camp "on duty," and after treating a patient sat reading outside my tent. My servant rushed up to me, and said that I was wanted to shoot a snake that had crept under some brushwood. I hastened to unpack my gun and ran after the boy, and soon came to where our men, in a state of great excitement, had formed a cordon round a patch of dry grass, to which they had set fire. Finally the snake came out, and all our fellows shouted at it. They were in mortal terror of it, as a fact, and certainly it looked "an ugly customer." It was too big to be stopped by shot unless I could make sure of hitting it in the head, and this I was not able to manage before it crawled under a saddle belonging to one of the soldiers. The saddle was lifted by the aid of a long pole, and in a moment the snake's head was smashed by the same means. I measured it, and found that it was just over two yards in length. The back was brown and the belly white, and the skull had the typical shape of the adder family. Generally speaking, we saw fewer reptiles in Abyssinia than we had expected.

After this incident, I strolled round the camp with my gun, and presently noticed a little grey animal scampering among the stones of a ruined house. It was of the same colour exactly as the stones, but presently I believed I could discern an eye, and, being anxious to ascertain what creature it was, took aim and fired. I walked up to the spot but found nothing, looked around and wondered how the animal could have vanished.

At the moment I heard the rustling of a leaf beyond the tumbledown wall on the left, and, guided by the sound, discovered the animal just dead. It was a specimen of the hyrax—an interesting creature to biologists, which Huxley described as "the type of a distinct order, in many respects intermediate between the *Ungulata*, on the one hand, and *Rodentia* and *Insectivora* on the other."* It is found only in Syria and Africa. I thought the skin worth preserving, and one of the soldiers flayed it for me.

On Sunday, February 8, our road lay, for the most part, at some little distance from the lake, which was out of sight till the end of the journey. We plodded on through long grass and past burnt patches. The track is only about a foot wide, and in consequence the loads of the donkeys extended beyond it on either side. When they were among the tall growth they had to sweep it aside from their burden as they went, and this tired them greatly.

We reached the edge of the lake at two o'clock, after an unbroken journey of six and a half hours, and pitched our camp. About two hundred yards from the shore five hippos were standing, well clear of the water. They looked for all the world like rocks, even when we brought field-glasses to bear on them. After lunch Crawley had the Berthon boat put together, and I rowed him towards the hippos. When we were within a hundred yards of them, I considered that the range was quite short enough to give the marksman an opportunity of displaying his skill, and he got no nearer. He began practice at once, but the boat was pitching rather sharply, and this made aiming with the rifle almost

* "The Anatomy of Vertebrated Animals," p. 367. This little animal is the coney of the Scriptures.

chance-work. Presently the sport became like firing at disappearing targets, for the hippos rose only once in two or three minutes to breathe. If they had taken my friend seriously, we should probably not have left Lake Tsana, and I felt relieved when he had had enough of the pastime and we rowed ashore.

Various small offerings of milk and bread reached us from the hamlets around. The milk is always sour. The Habashes do not drink it when it is fresh, and as a consequence they never wash the gourds in which they keep it, because it "turns" sooner in a dirty receptacle. At a few places we were able to have the cows brought into camp, and stored the milk in our own vessels, but this was impossible when we only remained one night on a camping-ground.

That evening Johannes, the interpreter, had a touch of fever. On the previous night his tent had been close to the papyrus swamp, and this, no doubt, accounted for the attack.

On the following day, February 9, our road lay through more broken country and more pleasing scenery. The track led us up hills, and down them, and between them, and sometimes close beside the lake. I saw no trees in this region but mimosas; the ground was covered in places by mimosa scrub, in others by dry grass. We made a march of seventeen miles, as we reckoned, a longer distance than we had travelled on the previous day. The donkeys were tired out at the end of the journey. Some stood still and refused to move, others lay down under their burdens in the path. We camped on the shore of the lake, at a spot very similar to that which we had chosen for our last halting-place.

The interpreter's fever had yielded considerably to

the usual quinine treatment, and he seemed very little the worse for his long ride in the heat.

In the evening the wind rose and presently blew hard enough to make me wonder whether my tent would collapse or not. I observed that very soon enough wash was knocked up in the deeper water to stop the headway of a rowing-boat. The lake would be a perfect place for fishing and sailing in the dry season. But without experience and watchfulness, risks would arise—quite apart from the humours of the hippopotami.*

On Tuesday, February 10, after a journey of about twelve miles through tall grass that impeded our baggage-train, we reached the village of Delgi, and pitched our camp upon the same ground which we had chosen for our first halting-place beside the lake. To the best of my belief, we were the only Europeans who had ever completed the circuit of this lovely reservoir of the Nile—the distance in all is one hundred and sixty to one hundred and seventy miles. Those who love regions beyond the outposts of our civilization, where the tourist ceases from troubling, could not seek isolation amid sweeter surroundings than this sunlit lake, these tropical mountains, and the quaint, quiet hamlets of a primitive people, who as yet, thank Heaven, have not been infected with "Yankee hustle."

The level of the surface of the water had sunk perceptibly since we were here a month ago, and many rocks were now visible which had then been covered. But I do not think the variation between the maximum

* Since we were by the lakeside a decree has been issued that the hippos, for their misdeeds in destroying crops and overturning *tankoas*, are to be treated as *vermin*, and shot whenever opportunity arises; so I fear that these ancient "riparian owners" will soon disappear from their pleasant and immemorial haunts.

height in the rainy season and the minimum in the dry is very great; I doubt if it exceeds eight or ten feet.

At midday we noticed that the climate into which we had come in the north-west corner of the lake was distinctly warmer than that of the other parts. We remembered the Soudan, towards which our faces were set, and knew that in the heat there we should think longingly of the waterside, in spite of all the worries inseparable from travelling in Abyssinia.

The "Sultan" of Delgi came to camp to welcome us as soon as he heard of our arrival. It was pleasant to meet this cheery, genial old Habash again, and his visit put us in good spirits. He brought us a couple of chickens and some bread, and we showed our hospitality by offering him chartreuse. He drank half a pint at a draught without "turning a hair," seemed none the worse for it, and wanted more. A man who remains sober after that performance must have a sound head and a strong body, but I do not recommend even the hardiest of my readers to attempt the feat. The Abyssinians who survive have inured their stomachs to excesses of abstinence from food as well as gluttonous excesses, an unwholesome and repulsive diet and the abuse of condiments, and they have constitutions *à toute épreuve.* I may mention that the children, who would be the first to suffer from bad climatic conditions, and the women seemed, as a body, fairly healthy, though they eat raw flesh like the men and suffer from the consequent parasitical trouble.

Nothing was mentioned or seen of the "Deputy" who had tried to prevent us from reaching the lake. But we made kindly inquiry for the "Sultan's" wife, who had shown us goodwill and sent us bread and *tedj* on our first arrival, and were glad to hear that the lady was

INTERVIEWING THE SULTAN OF DELGI. [*See p.* 170

THE SULTAN OF DELGI, AN OLD PRIEST, AND SLAVE CARRYING THE SHIELD. [*See p.* 170.

in good health. At nightfall we saw that the grass was blazing in two places, and it was interesting to observe with what extraordinary swiftness the flames ran over the ground when the breeze from the lake fanned them.

We had determined to show our liking for the "Sultan" by treating him handsomely when we gave the customary presents. He came into camp next day (February 11) as soon as we had finished breakfast, and we arranged the gifts on the table. They were a revolver, a folding-chair, a bottle of chartreuse, and a red silk cummerbund for his consort. He appeared to be well pleased and thanked us, through Johannes, with all the formalities of the country. We suggested that he should come for a row in the Berthon boat, but he backed out of this immediately. Then he had his new folding-chair carried to the shore, and sat there surrounded by his bodyguard, a force which would have had a great success in a pantomime. He followed our movements with much interest while we put the boat together, but he was plainly apprehensive when Crawley tested in still water the electric current-meter which he had used in calculating the outflow from the lake into the Blue Nile. It made a buzzing noise, and I think he suspected that it might "go off" at any moment. He remarked that it was *a watch* for the water, but he soon retired, followed by his comedians under arms.

Everything in camp that could be cleaned in water was washed in the lake during the day, and the ground near our tents looked like an improvised laundry. It was strewn in all directions with articles spread out to dry. Those boys who had no change of garments, stayed in the water while their clothes were aired in the sunshine, and enjoyed it.

Both Christians and Mussulmans received an ox on

this day in order that they might make biltong for the journey. We bought flour and grain and sheep and fowls, and laid in as large a stock of provisions as we could carry for use in the uninhabited borderland between Delgi and Gallabat.

February 12 was our last day beside Lake Tsana, and the prospect of leaving it filled me with regret. The "Sultan" came into camp again, and in answer to one or two questions gave us the following information: that the rainfall on the margin of the lake in the wet season is not great; the heavy rains descend upon the heights around, and the floods are carried down the rivers and khors in overwhelming torrents, so that all the watercourses are then impassable. This circumstance, of course, has a very important bearing on the possibility of a commercial development of the region.

Before quitting the subject, I will bring to the reader's notice Dr. Stecker's thoughtful conclusions as to the present relation of the River Abai to Lake Tsana:—

"I made another extremely interesting discovery in the Gorgora mountains, viz. of a remarkable shell, which by its character reminds one of the oyster.* We found both the shells and the living creatures in abundance on the shore. With lemon juice they taste like genuine oysters. But it is remarkable that I had already found this species in the Blue Nile, and that I found it later on the island of Dek, enclosed in unmistakable volcanic rock (tuff). I can only offer this explanation, that, at a time when Lake Tsana already existed, a great eruption took place in the south. According to my view the lake

* We found these shells and oysters in all the rivers to which our journey brought us. They are especially abundant in the shallow pools of the Atbara.

had its origin in tertiary times as a consequence of a great volcanic movement in the north (at the Gorgora range). The Abai, which was previously an unimportant stream, and described the great curve which is indicated on my map by arrows, and now carries it round Dek and Dega, was forced in consequence to the south-western and southern shores, though its original course can still be traced quite distinctly. The second volcanic movement took place, according to my opinion, in the south, and the islands of Dek * and Dega in Lake Tsana owe their origin to it, as do a whole series of islands beside the eastern shore of the lake and the masses of rock of volcanic origin which encumber the course of the Blue Nile and lie scattered in the whole valley of that river." †

The moon was full that night, and the lake and the

* I have already said that the island of Dek seems to me to have been chiefly formed by siltage.

† "Mittheilungen der Afrikanischen Gesellschaft in Deutschland," 1881, vol. iii. p. 28. The following general description of the lake district embodied by Mr. Dupuis in his Report on the Expedition, and published in Sir W. Garstin's Despatch already cited (Ed. 2165, 1904, p. 215), is terse and interesting:

"The depression occupied by the lake is, generally speaking, saucer-shaped, the land rising from the edge of the water in gentle undulating plains, getting steeper and steeper and terminating in considerable hills or mountains of some importance. In several places, as at Gorgora on the north, Mitraha and Koratsa on the east, Zegi on the south, and Dengelber on the west, the hills come right down to the lake and descend more or less abruptly into the water, but more often they stand back at some distance from it. The geological formation is almost everywhere of the same primitive character that seems to be universal in the Eastern Soudan, granite, gneiss, and quartz, varied only by tracts of lava, basalt, and eruptive rocks. Sandstone is reported by some travellers, and lime-stone is said to be found near Gondar, but I saw neither myself. The large tracts of comparatively level land consist almost entirely of the cracked black cotton soil usually found associated with igneous rocks. At the mouths of all the larger rivers flowing into the lake are extensive alluvial plains generally composed wholly of this same cotton soil, and of the greatest fertility, though nine-tenths of their area now grows nothing more valuable than coarse grass."

mountains formed a glorious scene, which left in the mind a longing to behold it again. Lake Tsana has that haunting, attractive power which some places possess, and which prompts one to return to them in spite of all that commonsense says about obstacles and discomfort.

On February 13 we made an early start upon our return journey to the Soudan, and reached the top of the plateau at half-past ten. We saw nothing, after all, of Hyli's men; I did not hear why they failed to join us. As we now retraced our steps over the same ground that we had traversed on the way to the lake, I shall not give any detailed account of our progress. In one day's journey we covered a distance which had given us an arduous two days' climb on the upward march, and we came again into the region of the Soudanese heat. The sequence of vegetation according to altitude was strikingly apparent during the descent. Throughout the first half of it we were in the cactus country, then in the bamboo zone, and, finally, among the mimosas and desert scrub. The heat proved trying to man and beast. We Europeans felt fagged and dull; I had three fever patients among our followers in the evening; and the donkeys straggled into camp jaded and spent.

An incident of the journey will show the nature of the road. A little while after we had passed the long and narrow gorge which has already been described, I caught sight of my valise and some other luggage stacked on a bank. Upon asking why it was there, I found that the donkey which had carried the load had slipped from the path and had fallen and rolled about ninety or a hundred feet. The beast seemed none the worse for the adventure, and the boys were bringing up the baggage piecemeal from the ledge which had stopped the donkey. I have no doubt that its burden saved it; a rider might have performed the same service if his remains had kept in place. A mirror and a candle-globe which "accomplished the descent" were not injured.

In spite of the steepness and roughness of the path—the mountains are genuine *sierras* (saws)—I believe that no great trouble and expense would be needed to make this approach to Western Abyssinia easy and safe, but it would be impossible for camels.

On the following day we reached the banks of the River Gerar, on which there is a thick growth of bamboos. Many of these had fallen across the track. We were not so much impeded by them as we should have been but for our previous experience of this part of the road. We sent men ahead with axes to clear the path as well as they could. In the evening we lost one of our baggage animals for the first time; two donkeys fell on the road exhausted. We left them while we moved on into camp, in the hope that they would recover. One was brought in later, but the other was found to be in such a helpless and pitiable state that nothing remained but to put it out of its misery, and it was shot. Our camp was beside the hot springs which have already been mentioned. We found in the neighbouring huts apparently the same company that we had seen on the upward journey. I was not able to learn anything about the people.

On the following day we left the valley of the Gerar, crossed a high ridge, and then followed the course of the Shemal Warhar. When we reached the place where we had pitched our camp before, we found the trees charred and the ground blackened by fire. There was no shade, and the water in the pools was very low. So we marched about three miles to another camping-ground called Ananta. Here water and shade were plentiful.

On February 16 we reached the bank of the Gundar Warhar. From midday till three o'clock the thermometer showed a temperature of 98° F. We had now entered the region in which robber bands are active. On the 17th we passed two parties of traders and heard from them our first news of the world outside Abyssinia. This was that Slatin Pasha had arrived at Gallabat on a tour of inspection. We thought of Gallabat—by contrast—as a centre of civilization. Our halting-place was a pleasant, shady spot beside the river-bed, in which water was abundant. In the afternoon Crawley shot a water-buck. I secured a much smaller prize in the form of another civet-cat. We heard that a large crocodile had been seen in a pool about two hundred yards from camp, but saw nothing of the beast. At night we were much plagued by mosquitoes and the beetles which swarm on one and do not bite, but are malodorous in death—a most perplexing pest.

On the 18th the temperature rose to 101° F. in the shade. We made a short march, but one of our donkeys succumbed and had to be left for the vultures. We found that the grass about our former camping-ground had been trodden down and a part of it burned. The water, too, was much lower in the pools, the dry season being now far advanced, though it was still abundant to meet all the requirements likely to arise in that country before the rains. At this time of the year the trees cast their leaves, so that there is much less shade along the track. On the 19th we reached the "warsha," which is the first from Gallabat on the road to Abyssinia and the last on the return journey. The ground was in a filthy state, and it was evident that many people had encamped at the place since we left it. Water was plentiful; it was drawn from a deep pool. To our horror, just as we had settled to rest in our chairs, we saw some of our boys and camp-followers washing their persons and their clothes in the water which we were to drink! One of them was a leprous trader. We raised some first-class trouble about this, and posted a sentry by the pool. But all we could do was to boil and filter the water thoroughly—and think of the other people who had bathed in it since we last tasted it. On Friday, February 20, we arrived at Gallabat.

I will not dilate on the traveller's delight in seeing again a batch of envelopes addressed to him in familiar handwriting. It is one of the pleasures which is becoming rare, but a trip into Western Abyssinia still provides an opportunity for it. In the town I met the Doctor of Kassala. He was then making a tour of inspection with the object of discovering cases of leprosy. The Egyptian Government has wisely ordained that lepers shall be compulsorily segregated; there is a hospital for them at Gallabat, which serves as a receiving station. An attempt is being made to form a colony of these stricken people, where they may cultivate the land and live by their labour.

The water which we had drunk at "Warsha No. 1" punished us rather severely for our rashness. And while suffering from this inward infirmity, I saw the camels that were waiting to wreak their eccentricities upon us during the remainder of the journey. It was our intention to follow the upper course of the river Atbara through Sofi to Kassala. This was formerly a frequented route, but the Dervishes had destroyed all the villages that stood beside it, and, as a consequence, the disused track had become impassable. It was then being repaired, but the work was still some way short of completion, and it seemed that we might be compelled, after all, to return as we had come by the road through Gedaref. However, our chief telegraphed to Mr. Flemming, asking him to send a gang of men, if possible, to the spot where the track was in the worst condition, and this he did. So we were able to carry out our plan, and had the satisfaction of being the first travellers who journeyed by the repaired highway from Gallabat to Kassala.

On February 22 we paid off the escort and followers who had gone into Abyssinia with us. Johannes received as presents a revolver and a watch, the others a gift of money. We found that nearly all preferred a settlement by means of drafts on Colonel Harrington at Addis Abbiba. If they carried cash through the "rain country" they would run a great risk of losing both their money and their lives. Before we bade farewell to Johannes, he told us that we had on one occasion, unconsciously, been in no small danger in Abyssinia. At a certain village a rumour was spread that King Menelek had sold the lake to the British. The supposed transfer was by no means to the taste of the Habashes—and we represented the English interest. Disregarding the niceties of French grammar, Johannes remarked that he had *beaucoup de peur* at the time. In our case ignorance was bliss.

On this day we said good-bye to Zodi. He had decided to return to his own country with Johannes. We all wished that he could have finished the journey with us, but felt that we should not be justified in discharging him finally in a country where he was ignorant of the language. He took his leave of us, bowing to the ground and addressing us in his own tongue. His face bore an expression which it is difficult to describe; he was showing his spotlessly white teeth in a smile, but he was half crying nevertheless. I have always wished that I could have engaged him as my servant. He was an excellent and faithful lad, and I believe that he did us valuable service when we were near Zegi, and the Fituari's son was inclined to give trouble. Zody, as a native of the lakeside (Korata), was able to influence his neighbours on the opposite shore when a Habash from another district might have been disregarded or set at naught.

My two companions went out with their rifles in the afternoon, and there was no slight commotion in Gallabat when they were heard firing at antelopes in the distance. During our absence, as I have already mentioned, an Abyssinian band had raided two villages and carried off thirty-eight men and boys into slavery. The noise of the discharges made the natives think that the slave-raiders were at their work again. The very serious political consequences which these incursions will certainly cause sooner or later have been alluded to before.

It may interest my readers, if, before saying the last word about Gallabat, I give a brief account of the battle which took place there in March, 1889, between King John of Abyssinia and the Dervishes. Perhaps no incident in modern history so strangely combines the oriental and the mediæval atmospheres, or so oddly illustrates the effect of weapons of precision in warfare of the biblical type. A detailed narrative has been given by Mr. Augustus Wylde.*

The Dervishes had invaded Abyssinia in 1887, at a time when the forces of the Negus were scattered. They defeated Tecla Haimanot and devastated the region to the north of Lake Tsana. In 1888 they renewed the invasion, but in the meanwhile the King of Godjam's army had been reinforced and his troops had been

* "Modern Abyssinia," pp. 40–43.

rearmed. The Dervishes suffered a severe defeat. The Negus completed his preparations in the winter of 1888–9, and gave notice to the Khalifa of his intention to march upon Omdurman. The Dervishes massed about seventy thousand men at Gallabat, where they occupied a large zareba which was protected by a dyke and some redoubts. They also had artillery in the fort where the scorpions have now become so formidable. But they were badly supplied with small-arms, whereas King John's men had a plentiful stock of effective rifles of a French pattern.

The Dervish position was surrounded by the Abyssinian army. The Khalifa's men were crowded in their enclosure, and, owing to its construction, could not fire from it without exposing themselves. The Abyssinian marksmen did terrible execution, and finally a body of King John's horse, supported by a hot rifle-fire at short range, burned the thorns of the zareba at several points and filled up the ditch. The position was then "rushed," and only a few of the Dervishes, aided by the smoke and the confusion, escaped.

"Facing King Johannes's bodyguard," says Mr. Wylde, "one small redoubt, strongly fortified and held by the black slave soldiers of the Dervishes, still held out, and their rifle-fire was doing some execution. The King, getting angry that it had not been taken in the rear by the troops that had entered the sides of the fortifications, and who were engaged in plundering, went forward to attack it with his followers. The gaudy dresses worn by his staff, with their silver shields and the bright silks, drew the fire of the defenders. King Johannes was struck by a bullet that traversed the lower part of his arm and entered the intestines near the navel, taking into the wound a part of his dress. He still gave orders, and kept on the field till the redoubt was rushed and those in it all killed."

The King died about twenty-four hours after he was wounded. Quarrels with regard to the succession immediately commenced among the Rases, and instead of following up their victory, they retired at once into Abyssinia with their captives and plunder, in order to serve the interests of their respective factions.

"On the 11th, in the afternoon, old Ras Areya, the King's uncle, a man nearly eighty years of age, who had played a wonderful part in Abyssinian history, was left with a few followers to bring back the King's body for burial. The body had been cut in half so that it could be carried more easily, and was put in a clothes-box so that it could be laden on a mule. Only a few of the King's devoted servants remained behind, with a few priests and their armed servants. On the 12th, while following the Taccazze road, the sad and mournful procession was overtaken by a few Dervishes and some Arabs, who had returned on the night of the 10th to reconnoitre Gallabat, and when they found it abandoned they had followed one of the lines of retreat to find out what was going on and the reason the Abyssinian victory had not been followed up." *

Ras Areya could have escaped, but died, with a few of his soldiers and the bravest of the King's servants, defending all that remained of his liege. He "was last seen standing alongside the box containing the King's body, after having expended all his ammunition, with his shield and sword in his hands." He was speared by a Dervish from behind. The Khalifa made no attempt to invade Abyssinia in force after the battle of Gallabat.

* "Modern Abyssinia," p. 42.

Mr. Wylde received this information from a priest who was present and who escaped, though he was badly wounded. I was told that the King died in a small hut, attended by two monks and four nuns, who tried to keep his death secret, and that an old woman, a servant of the nuns, being captured later by the Dervishes, revealed, in order to save her life, the line of retreat of the party that was bearing away the King's body.

The encounter took place at night, and such a scene in the moonlight seems more like an imagined passage in an epic than an occurrence in the closing years of the last century. I do not know whether the nuns accompanied Ras Areya, nor what their fate was. The Dervishes cut off the King's head and carried it to the Khalifa at Omdurman.

There are numerous graves around Gallabat of those who fell in the engagement. These are mounds, on which are laid agates brought from the bed of the Atbara. The tombs of the Soudanese have upon them a calabash, or more usually a pair, containing grain and water—presumably to satisfy the needs of the dead man's shade. I do not know how the custom of solacing the deceased thus is reconciled with the doctrine of Islam.

CHAPTER XII

The valley of the River Atbara can now scarcely be regarded as unfamiliar country. Those who wish to read a most lively, interesting account of all the opportunities that it offers to the sportsman, should turn to Sir Samuel Baker's book, "The Nile Tributaries of Abyssinia." Our journey from Gallabat to Cairo by way of Kassala can be quite briefly described.

In reply to our inquiry we received a telegram from Mr. Flemming, who told us that a guide would meet our party at an appointed spot to show us the now disused and obscured road, and on the morning of February 23 we left Gallabat. About four hours' travelling over flat, rocky country brought us, by an easy descent, to the Atbara. The banks here are low, and the bed is hard and clean, with tracts of shingle in it. The breadth of the course is about a hundred and fifty yards, and there were large pools of clear, good water. Many of them were deep. At this season of the year an insignificant stream trickles from pool to pool, and sometimes disappears altogether. There are big fish in these pools, and they need cautious handling. I took my tackle after we had pitched our camp, landed a two-pounder, and was "wool-gathering" a few moments afterwards when a powerful fish seized my bait with a dash and had me off my balance in an instant; result, a bruised knee and two hooks lost.

Next day Dupuis moved ahead of the party with his Soudanese "Shikari," and in a couple of hours stalked and shot a gazelle and a bushbuck with a fine pair of horns; later in the day a haartebeest fell to his rifle. We saw great numbers of gazelles and ariels, and the country was full of game. The tracks left by many kinds of animals going to and from the river were visible in all directions. In addition, guinea-fowl were plentiful. The ground near the river was covered with dry grass, and there was a thin but continuous growth of mimosa scrub. The baobab tree flourishes in this region.* We camped on the site of a village that had been laid waste by the Dervishes. No vestige of it remained but potsherds.

When the flesh of the animals was being cut into strips for use as biltong, a host of the carrion birds of all orders gathered around us—crows, hawks, vultures, and marabous. There was something disquieting and unpleasant in the presence of these groups of eagerly expectant scavengers, which faced us wherever we turned, and eyed us from every tree. That night the mosquitoes "rushed" my curtains and made a most successful raid, retreating at sunrise from the stricken —and bitten—field.

Shortly after we had started on February 25, I saw, at a little distance from camp, a piece of neatly made basket-work, and picked it up. Further on I found another. On

* *Adansonia digitata.* These are tall trees, whose timber is of a spongy consistency. Sir S. Baker mentions one which was about forty feet in circumference. He wrote: "The *Adansonia digitata*, although a tree, always reminds me of a gigantic fungus; the stem is disproportioned in its immense thickness to its height, and its branches are few in number, and as massive in character as the stem. The wood is not much firmer in substance than cork, and is as succulent as a carrot." ("The Nile Tributaries of Abyssinia," p. 356.)

HEAD OF A HAARTEBEEST. [*See p.* 184.

'LATES NILOTICUS' CAUGHT WITH A TROUT-ROD.
[*See p.* 200.

making inquiry I was told that these were parts of game-traps. A log is buried in the ground, and a piece of cord with a noose at the loose end is attached to it. Then a hole is dug where tracks of game are seen, the noose is "set" in the hole, the basketwork is laid down to cover the trap, and the cord is hidden by a layer of earth. When a beast steps into the hole, the movement of the basketwork pulls the noose tight around its leg. The more the beast struggles, the firmer the grip becomes; finally, as a rule, the log is pulled up by its exertions, and the hobbled creature limps away, but is easily caught, and can then be killed as the Mohammedan rite requires.

The track in this region is kept open by the cutting down of mimosa scrub. It lies upon "cotton soil" with fissures in it, large and small. The camels, which stare stupidly ahead while they walk, are constantly stumbling on account of these holes; their legs seem to have a marvellous instinct for finding them. Another difficulty was the descent and ascent in numerous little khors which run down to the river bed, and the camel is singularly clumsy when it leaves level ground. We pitched our camp at midday on the site of another deserted village, called Wad Abou Simam. Before that time each of my friends had shot a waterbuck. Very sensibly, the Anglo-Egyptian Administration imposes a fine on those who shoot the females of the larger antelopes.

I think we must have seen hundreds of thousands of guinea-fowl during the last two days' journey. They were in the jungle of mimosa scrub and neblik and on the shingle in the river bed, where by reason of their colour they can scarcely be distinguished from the stones; in fact, they swarmed in all directions. I took my gun into the dry course of the Atbara and sat down near an islet which was covered with very dry grass. This was a favourite place of refuge for the guinea-fowl. Dupuis drove them to the bank, where they rose, and flocks of them flew towards the little island. In this way I had some fine practice, and the birds were always useful in the larder. Travellers in this region should set up their mosquito-curtains with care. I was remiss about it that night, and the determined and virulent insects bit my scalp, face, and hands through the covering. The next evening I was more attentive to the matter, and formed a frame by means of four dhurra stalks. This arrangement, with draping, had all the advantages and some of the solid dignity of a four-post bedstead.

On the following day (February 26) we encamped at Sherafa. There was no incident worthy of mention during the journey. In the afternoon Dupuis and I saw a crocodile within range in the river bed, and fired at the same moment. The only visible result was that he scuttled into a pool and disappeared.

On the 27th Crawley brought down an ariel, but we had entered a district in which game was much rarer. We reached a well at a place called Tubra Cullah. This has been dug at the edge of a rock, the water was brackish, and there was no trough for the camels. We had now left the course of the Atbara, but had, fortunately, brought with us water from the pools which was sufficient for the men. On this day the temperature rose to 107° F. in the shade. The road, which had at that time been cleared, ended at Tubra Cullah, and here we were met by the guide whom Mr. Flemming had sent to show us the unrepaired track between this place and Sofi.

HADENDOWA, CAMEL DRIVER. [*See p.* 186.

DINKA BOY, CAMEL DRIVER. [*See p.* 186.

On the following day we marched to Goratia. Our guide had misinformed us as to the distance, which was much greater than we had expected. The heat was scarcely endurable, and none of our marches had been more fatiguing. How we longed for the coolness and freshness of the lakeside! Close to the village was a well with good water and troughs for the camels, which had not drunk for nearly forty-eight hours. We found the inhabitants very willing to help our men. While we were passing through Goratia the women within the huts warbled their welcome in the usual manner and with unmistakable enthusiasm. We learned that when our party was first sighted, a rumour spread that we were Abyssinian raiders, and the non-combatants rushed to the shelter of their homes. The relief of the villagers was evidently very great when they found that we were English people.

The following instance may serve to show the difficulty which British administrators have in suppressing the slave traffic.

A certain man, a native of the White Nile region, appeared one morning before Colonel G. at Rosaires and said that while he was taking a convoy of slaves into Abyssinia, he had been stopped by the sheikh of a village in the Colonel's district. This sheikh, he asserted, took possession of the slaves, and bound him and his companion, to whom they belonged. In order to keep what he had captured and destroy those who could bring him to justice, the sheikh had the two slave-dealers brought to a precipice and cast over it. One was killed, the other escaped with whole limbs. Then he repaired to the Governor's quarters to report the matter.

Colonel G. ordered the sheikh to attend and bring the slaves that he had taken. The sheikh, astounded to hear that the Moudir * had learned what he had done, attributed his misfortune to the Will of God with the usual phrase "Inshallah," † and started for the Moudirieh accompanied by the slaves. On the way he consoled himself with the words, "God will help me" (*Rabonna effrighi*).

While he was upon the journey he met two native merchants returning to their village, and immediately he concluded that Allah had heard his prayer. "Um del Allah!" said he, "God is good." He caused the merchants to be bound, and then he had the slaves brought before him, and made them all swear upon the Koran that these two men were they who had raided the village and carried the people away captive. And the merchants were forced to follow him with the slaves.

When the sheikh arrived at the Moudirieh, the Governor ordered him to deliver up the men in whose possession he had found the slaves—and the sheikh forthwith had the two merchants led forward. Colonel G. was in a dilemma. Evidently the raider who had complained to him was telling the truth, but all the slaves swore that they knew nothing of the man. Questioning was of no avail, and did not shake their evidence. Then the Moudir, turning suddenly to a little girl, asked her, "Who told you to say that these two accused men raided your village and took you away?" The little girl instantly pointed to the sheikh and answered, "Please, sir, that man."

This decided the matter. Some time after, the body of the slave-dealer's companion was found. I did not hear what befell the sheikh, but the Moudir was not a man to be trifled with, and I have no doubt that justice was done.

The sheikh of Goratia came into camp to greet us,

* Governor. † "If God is willing."

and afterwards sent us hot coffee, which was excellent. It is the most refreshing drink of all after a tiring journey in the baking sunshine.

In this village we found a man who declared that he could act as guide to the junction of the Atbara and the River Salaam, a point which is of interest in connection with the all-important problem of water-storage and distribution. It has not been sketched or mapped—indeed, I believe that no European has yet succeeded in reaching it, though the British officers in the Soudan have been eager to discover the spot. My two companions arranged to start on the following morning and endeavour to make a rough survey of the place.

At Goratia I obtained a photograph which shows clearly the amulet worn by almost all the Soudanese. The youth who wore it had been married about a week, and the scars on his back illustrate a singular custom among the people. Sir Samuel Baker has described it.

"There is but little lovemaking among the Arabs. The affair of matrimony usually commences by a present to the father of the girl, which, if accepted, is followed by a similar advance to the girl herself, and the arrangement is completed. All the friends of both parties are called together for the wedding; pistols and guns are fired off, if possessed. There is much feasting, and the unfortunate bridegroom undergoes the ordeal of whipping by the relations of his bride, in order to test his courage. Sometimes this punishment is exceedingly severe, being inflicted with the coorbash or whip of hippopotamus hide, which is cracked vigorously about his ribs and back. If the happy husband wishes to be considered a man worth having, he must receive the chastisement with an expression of enjoyment; in which case the crowds of women in admiration again raise their thrilling cry. After the rejoicings of the day are over, the bride is led in the evening to the residence of her husband, while a beating of drums and strumming of guitars (rhababas) are kept up for some hours during the night, with the usual discordant idea of singing." *

It is interesting to compare Mansfield Parkyns's description of the mode of duelling in Dongola.

"This duel is not a serious matter, but is engaged in by young men on the slightest possible pretext, often merely to display their manhood. An angareeb, under which a jar of beer is usually set (for refreshment between the rounds), is placed between the combatants, each of whom is stripped to the waist, and armed with one of the formidable whips I have described" (coorbash). "As soon as all is ready they begin, giving alternate stripes on each other's shoulders, but neither being allowed to evade or ward off a blow; they continue this agreeable pastime for a very long time, till one falls down exhausted from loss of blood and the punishment he has received." †

I spent a lazy day on March 1, watching the peaceful, patriarchal life of the village, and especially the flocks and herds when they came to the well to drink. Men, women, and boys all helped to fill the cattle-troughs, which are made of baked mud. The surroundings described in the book of Genesis seemed to have been preserved here without change of any kind.

My friends did not return on March 2, as they had expected. I took my rifle and a guide and tried to bring home an ariel, but could not approach within three hundred yards of the game. At about that distance one was hit, but got away. On the 3rd the guide and I went

* "The Nile Tributaries of Abyssinia," pp. 124, 125.
† "Life in Abyssinia," pp. 385, 386.

SOUDANESE WITH AMULETS. [See p. 189,

AT THE WELL, GORATIA. [See p. 190,

in another direction. Creeping carefully up against the wind, I aimed at a buck at three hundred yards' distance. He fell forward, burying his horns in the ground, when I fired, but was up again, and ran in a circle and then fell once more. I rushed forward to give him the *coup-de-grace*, but he bolted over the brow of a hill at the top of his speed, and we lost sight of him. I wandered about seeking for his tracks, but could not find them in that tumbled country. Later I saw three more ariels, which ran into some dry grass. We made a circuit, and presently came within sight of them again. They were evidently on the alert. Under cover of a tree, I took a steady aim and hit a buck, who limped away and lay down, but presently rose at eighty yards distance as we advanced, and I then killed him by a shot through the heart. We covered the body with branches to protect it from vultures, and carried the head into camp. I kept the skin and the head, but gave the flesh to my guide, who was much delighted. He went out on a donkey and cut the beast up, and distributed presents of meat among his friends.

My companions returned half an hour after midday. They had had difficulty in finding the course of the Atbara, and their camels had been on the march eight hours when they reached it at a gorge in coarse sandstone rock, where the river forces its way through a narrow cleft about twenty metres wide, between cliffs rising vertically from a profoundly deep pool. There they determined to encamp and spend the next day in attempting to find the junction of the rivers. To the great disappointment of all, they failed to achieve their purpose. The country in which they were is extremely wild. A lion had killed one of their camels in broad daylight. The boy who was in charge of the beasts while they browsed saw the lion creep from the thick undergrowth, spring upon the camel and crush its skull with one blow of the paw. He screamed, and the lion walked off into the scrub in which it had been concealed. The other camels were terrified, and would certainly have bolted in a panic if they had not been hobbled. At night my friends heard hyenas running within a dozen yards of their beds.

They had been much interested in watching the huge flocks of small birds of the linnet kind which assembled half an hour before sunset and went to drink together in the pools of the Atbara. They come with an undulating flight, and, small as they are, the rushing of the wind as they beat the air makes a noise like thunder, and their numbers darken the sky. The weight of the throngs of them which alight at a time bends down the ends of the overhanging branches and twigs to the level of the water. I had seen these flocks many times, and the cunning trick which the crocodile uses in preying upon them. This was a point of Soudanese natural history which did not escape the observation of Sir Samuel Baker.

"Few creatures are so sly and wary as the crocodile. I watch them continually as they attack the dense flocks of small birds that throng the bushes at the water's edge. These birds are perfectly aware of the danger, and they fly from the attack if possible. The crocodile then quietly and innocently lies upon the surface, as though it had appeared quite by an accident; it thus attracts the attention of the birds, and it slowly sails away to a considerable distance, exposed to their view. The birds, thus beguiled by the deceiver, believe that the danger is removed, and they again flock to the bush, and once more dip their thirsty beaks into the stream.

Thus absorbed in slaking their thirst, they do not observe that their enemy is no longer on the surface. A sudden splash, followed by a huge pair of jaws beneath the bush that engulfs some dozens of victims, is the signal unexpectedly given of the crocodile's return, who has thus slily dived, and hastened under cover of water to his victims. I have seen the crocodiles repeat this manœuvre constantly, they deceive by a feigned retreat, and then attack from below." *

These birds fall a prey not only to crocodiles but to large fish—of what species I am not sure—which rise at them as other kinds rise at flies and snap them off the twigs. Besides, kites scout about the outskirts of the flocks as they fly and pick up the stragglers.

In this part of the country we saw baobab trees only on the banks of the river. Elsewhere, the trees are all mimosas and the undergrowth of the same order.

On the 4th we marched over a dead level of cotton soil to an abandoned village. We had to carry drinking-water, for the well at this place had been filled up by the Dervishes. I have very little doubt that this region is full of malaria in the wet season. Our camp was at the foot of a rocky hill, on which we found abundance of guinea-fowl when we took out our guns late in the afternoon. On the following day we saw herd after herd of ariel as we marched. They were extremely tame, and we passed within ten yards of some of them. The track which we followed brought us to the Atbara again, at the hamlet of Aradeeb. This village was poverty-stricken and almost in a state of famine. The aphis †

* "The Nile Tributaries of Abyssinia," p. 240.

† This blight is known as "asal," or "honey," from the sticky deposit formed all over the plant and leaves. At the time of the expedition it prevailed more or less severely throughout the greater part of the Kassala Moudirieh. The stalks of the plants attacked become black.

had blighted the entire dhurra crop of the inhabitants. I could not discover how the people lived, but they seemed to bear their trouble uncomplainingly, and here as elsewhere appeared to be sincerely thankful for the peace and safety which the Anglo-Egyptian rule secures to them. In every village to which we came we heard the same remark—"Miri quies," "the Government is good." I believe that this expression of opinion is perfectly sincere. If so, it is one of the greatest of the many great triumphs of British administration.

The course of the Atbara is much narrower here than at the point where we had left it, and the pools are deeper. They are not connected by a current in the dry season. I judged that the stream would be from twenty-five to thirty feet deep when the floods are at their height. The river runs, with a shingly bed, through a curious formation of coarse, gritty sandstone, which forms fine cliffs and numerous rocky bars and barriers. The sandstone stratum extends to about two-thirds of the height of the ravines, and the upper third consists of the usual black-cotton soil.

When the heat of the day was over I caught a couple of fish of about three pounds and four pounds weight. They showed some sport and made very good eating. In the evening we should have rested in perfect contentment, after enjoying the luxury of a bath, if insects of various sorts had not swarmed upon us incessantly, crawling and biting. Dupuis gave the sheikh of the village two sheep which we had brought from Zegi. They suffered severely from the heat, and it seemed cruel to drive them further. Besides, we were glad to add something to the store of the villagers in their time of scarcity.

On March 6 we quitted Aradeeb. The old sheikh

came to our camp to see us start, and brought us hot coffee in which we ceremonially drank "peace to the village." Throughout our journey on this day we followed the course of the river, constantly crossing gorges and khors. We reached Sofi at midday. This is the large village near which Sir Samuel Baker dwelt for five months in 1861. He has given a very full and most interesting account of the surrounding district in "The Nile Tributaries of Abyssinia." There are rest-huts built of straw at Sofi, and we were glad to take refuge in them, though the temperature within when we arrived was 107° F. However, we were able to remove our helmets, and this we could not do in our tents in the daytime as the sun penetrated the canvas. At night the temperature was 98° F., and a hot wind was blowing strongly. Towards evening I had taken my rod to a pool in the river, but this unpleasant breeze had by that time ruffled the surface smartly, and I got no sport. I think the fish lie quietly in shelter during these storms and do not feed.

Next morning we left Sofi, passed through the village of Tomat, and about midday pitched our camp five miles beyond this place opposite to the "mugra," that is the junction of the Settite and Atbara rivers. The two streams at the point of meeting have a course about a quarter of a mile broad. The bed of the Atbara is flat and there is much shingle in it. The sandstone formation has come to an end, and the rocks which crop up here and there are of a granitic character. Very little water was flowing in either river, but a larger stream trickled in the Settite than in the other. We did not stir far from camp in the afternoon, but shot sand-grouse as they came to drink in the pools.

During the day we saw tracks of lions, hyenas, and

monkeys—grivets and baboons. The soil and the vegetation by the river were such as we had seen on the previous marches. A lion came into the camp during the night. We were sleeping in the open and heard him sniffing round our beds. I sang out to Dupuis, and asked him what was making the noise. He struck a light, and the beast skulked off into the bush. My two friends snatched up their rifles and hurried off to see if he meant to attack the camels. That, no doubt, had been his errand, for he left us alone, though I saw his tracks within a yard of my bed next morning. The camels, as usual, were gathered in circles near the men's fires. They were hobbled, and I heard them stamping quickly on the ground in their fright, making a strange pattering noise. I believe they only show fear in this way when they smell a lion.

The next day we marched to Khor Katout, a big ravine in the plain of the Atbara, which extends to the river. A hot wind had again arisen, and was carrying quantities of dust with it. The journey was exceedingly unpleasant, and coolness was nowhere to be found. Khor Katout is a well-known haunt of lions, and we heard them roaring in the night. On March 10 we arrived at the point where the road from Gedaref to Kassala joins that on which we were travelling. Here there is a "nocter"—a military post, where five soldiers are stationed. At this place we found a rest-hut, in which we took shelter till three in the afternoon, and then marched on till six, covering about nine miles in the three hours. Since we had left the course of the Atbara we had been obliged to send to the river for water. It was brought in the "fantassis."* This necessity delayed

* These are watertanks of galvanized iron. Two of them form a camel's load.

us, and added to the petty troubles of the journey, but we expected to reach the river bed again on the next day. That night my sleeping-place in the open was within three or four yards of the track, and the moon was very bright. It was a most characteristic Eastern sight, when, about nine o'clock, some thirty camels passed along the route at the foot of my bed, slowly and silently. They carried no loads, and were probably returning from Kassala to Gedaref.

On the following day, as we marched, we saw spoor of lions and leopards among the mingled traces of hyenas, jackals, and many antelopes. We camped for the night at a very lonely spot on the Atbara which is called Khashim el Girba.* Here the river emerges from a narrow, deep, rocky gorge where the width of its course in places does not exceed one hundred and twenty feet. The steep sides have been smoothed by the terrific rush of water in the rainy season. One of the pools in the bed of this cataract is very large, and uniformly deep. I noticed that earth had lodged behind rocks

* Mr. Dupuis's account of this place has been published in Sir W. Garstin's Report (p. 219), and is in the following terms: "This is a very remarkable spot, and the probable site of large canal works in the future, if such are ever undertaken on the Atbara. Just above the "meshra," or watering-place, the Atbara flows in a narrow deep trench, perfectly straight for a couple of miles or so, with a width of little over one hundred metres, and a depth of about ten metres all across the pool when at summer level. The sides of this trench are nearly vertical cliffs of extremely hard granitic rock rising twenty metres or more from the water. At the meshra this trench suddenly fans out into a wide shallow channel, the pool ceases, and the river breaks up into three or four separate branches, amidst a confusion of rocky islands, and is not fully re-united into a single stream for several miles. The rocky substratum through which the river has cut its way here gives little or no indication of its presence in the level uplands, which continue as before to form an unbroken bush-covered plain of black soil up to within a couple of miles or so of the river, where it suddenly breaks into ravines."

which jutted from the sides of the gorge, and a growth of mimosa scrub had established itself even in the shallow soil on these ledges. There is no village at Khashim el Girba, but it is a recognized camping-ground. Beyond the narrows the Atbara broadens to a width of half a mile, and its course is divided by an island. Here, at the time of our visit, one could wade across the river bed through a stream which reached the ankles. The scene must be one of singular beauty and grandeur when the floods are coming down from Abyssinia. We heard a lion roaring at night, and there are numbers around this place, but they had plenty of game to prey upon and did not approach the camp.

On the morning of March 12 the Berthon boat was put together, and my companions took her, with some surveying gear, to the great pool in the gorge. I followed and carried my angling-tackle. We rowed about half a mile up the course of the river upon this stretch of water, and then came to the narrowest part of the bed of the rapids. Here a very large rock that has an accessible side juts out into the pool, in a most convenient position for survey work, and my companions began to take measurements, walking up a gravelly incline which seemed to have been placed there for the purpose. Their object was to mark out a gauge which would show the rise and fall of the river during the flood season in this gorge, where, of course, the movement is very clearly seen. My friends made the necessary marks in chalk at the time, and these were afterwards chiselled in the rock by a mason, who was sent down for the purpose. The gauge has since been connected by electric wire with Cairo, and since then in flood time the variations in the depth of the Atbara have been telegraphed every other day to the

Irrigation Department in the capital. This has proved to be a very satisfactory arrangement. If an exceptionally heavy flood occurs, the news is sent to Cairo immediately; warnings are then dispatched up and down the Nile so that all concerned may be prepared for the consequent rise of the river, which will take place, say, in six weeks' time. The Irrigation Department and the landowners look to their culverts and strengthen weak parts of the banks, and all is in readiness when the flood comes. Formerly it was the custom to patrol the riverside night and day when an unusual rush of water was expected, but the risk of disaster in Lower Egypt from this cause has been almost ended by the operation of the dam at Assuan and the precautions now taken.

I tried my luck in the pool, where I saw many huge fish, but they did not bite, though a number smelled at the spoon bait and some struck at it with their tails. I made attempts with other attractions, but the copper spoon appeared to allure them most. Then I cast the line into several pools in vain, but presently, in the shallower part of the big one, hooked a fish of about four pounds' weight, which fought hard, lashing the water, and plunging and keeping the reel busy. It took me half an hour to play him into a shallow, and in the meanwhile many big fish came up to see what was making the disturbance.

I went back to the deeper part of the pool, and as soon as I had made a cast, hooked one, whose first tug showed that he was a prize worth landing. I feared that I should lose him, as my tackle was hardly strong enough for a heavy catch, so I made up my mind to play him very gently and tire him out, if possible. Away he went and round spun the wheel. He hid under a ledge of rock, and I wound it gently. This started him again, and now he showed sport without a break for twenty minutes. After that, to my surprise, I brought him up quite close to a shallow, and then I saw that I had hooked a real monster and had a good chance of landing him. But it would be useless to try to get him ashore without help, and I shouted and shouted again; for all the time he was making dashes and stopping in exhaustion, and I was playing him in nearer to the shallow. At last one of our boys came to aid me, and I saw that he was eager to help—but the first thing that he did was to rush into the water and grab my line! I yelled to him to leave it alone and strove to explain what he was to do. The next time my prize came into the shallow his strength had nearly failed him, for he rolled over and showed his white belly. It was a moment of excitement when the black fellow slipped into the water and tried to catch hold of *the fish's tail.* I did nearly lose him then, but he was almost spent; and after a great deal of shouting I made the boy understand at last that he was to put his thumbs into the fish's gills on each side. At first he thought I wanted him to put his fingers into its mouth, and was afraid. It took two men to lift my fish on to dry ground, and I could hardly believe that the tackle had held him; I was using a small trout-rod. From snout to tail he measured forty-five inches, and he had a girth of thirty inches. We judged his weight to be between fifty and sixty pounds. This species (*lates niloticus*) is called by the natives "el baggar" (the cow). It has been described by Sir S. Baker, who also published in "The Nile Tributaries of Abyssinia" a drawing of a fine specimen which he caught in the Atbara. My photograph shows that the fish which I landed does not altogether correspond to his description. It may be that the Soudanese name covers more than one species, or there may be considerable variation among members of the same species, but I am inclined to think that the illustration in Sir S. Baker's book was prepared from an imperfect sketch.

During the morning we saw a troop of baboons. For a time they watched us. It was interesting to note their method of scouting. Here and there one would climb a tree, scan the surroundings sharply, and then descend quietly and join the main body. A few minutes later another would renew the observation from another tree. In this manner they keep a constant look-out. Mansfield Parkyns studied the habits of these creatures in Abyssinia and has given a most interesting account of them.

"The monkeys, especially the Cynocephali, who are astonishingly clever fellows, have their chiefs, whom they obey implicitly, and a regular system of tactics in war, pillaging expeditions, robbing cornfields, etc. These monkey forays are managed with the utmost regularity and precaution. A tribe, coming down to feed from their village on the mountain (usually a cleft in the face of some cliff), brings with it all its members, male and female, old and young. Some, the elders of the tribe, distinguishable by the quantity of mane which covers their shoulders, like a lion's, take the lead, peering cautiously over each precipice before they descend, and climbing to the top of every rock or stone which may afford them a better view of the road before them. Others have their posts as scouts on the flanks or rear; and all fulfil their duties with the utmost vigilance, calling out at times, apparently to keep order among the motley pack which forms the main body, or to give notice of the approach of any real or imagined danger. Their tones of voice on these occasions are so distinctly varied that a person much accustomed to watch their movements will at length fancy—and perhaps with some truth—that he can understand their signals.

"The main body is composed of females, inexperienced males, and the young people of the tribe. Those of the females who have small children carry them on their back. Unlike the dignified march of the leaders, the rabble go along in a most disorderly manner, trotting on and chattering, without taking the least heed of anything, apparently confiding in the vigilance of their scouts. Here a few of the youth linger behind to pick the berries off some tree, but not long, for the rearguard coming up forces them to regain their places. There a matron pauses for a moment to suckle her offspring, and, not to lose time, dresses its hair while it is taking its meal. Another young lady, probably excited by jealousy or by some sneering look or word, pulls an ugly mouth at her neighbour, and then, uttering a shrill squeal highly expressive of rage, vindictively snatches at her rival's leg with her hand, and gives her perhaps a bite in the hind-quarters. This provokes a retort, and a most unladylike quarrel ensues, till a loud bark of command from one of the chiefs calls them to order. A single cry of alarm makes them all halt and remain on the *qui vive*, till another bark in a different tone reassures them, and they then proceed on their march.

"Arrived at the cornfields, the scouts take their position on the eminences all round, while the remainder of the tribe collect provision with the utmost expedition, filling their cheek-pouches as full as they can hold, and then tucking the heads of corn under their armpits. Now, unless there be a partition of the collected spoil, how do the scouts feed? for I have watched them several times, and never observed them to quit for a moment their post of duty till it was time for the tribe to return, or till some indication of danger induced them to take flight. They show also the same sagacity in searching for water, discovering at once the places where it is most readily found in the sand, and then digging for it with their hands, just as men would, relieving one another in the work if the quantity of sand to be removed be considerable." *

In the afternoon Crawley shot a wild pig. I walked to the river again, being charmed by the lovely scenery, and as I approached saw numbers of crocodiles scuttle into the water. None was of great size. I doubt if any region in the world affords more varied sport than the valley of the Atbara. The climate is perfectly healthy in the dry season.

On March 13 we struck our camp in the afternoon and soon met the first European whom we had encountered since we left Gallabat. He was a British officer on the way from Kassala to Gedaref, and he made mention of the slave-raids which had taken place since we left the latter place. It seemed likely that the Arab Battalion would be kept busy in holding the marauders in check. In the course of our march we crossed the Atbara, at a point where it is about four hundred yards in breadth. Water that reached our ankles was trickling among the shingle in places. At sundown we arrived at a camping-ground called Fashur, and halted for the night. Our dinner was "bully beef" and pickles — neither a luxurious nor a prudent meal. We were now suffering from heat eczema, which is most irritating in the cool

* "Life in Abyssinia," pp. 102, 103.

of the night. The baying of a hyena kept me awake for hours, and a bed in the Soudan is not a pleasant place for one who lies on it with an itching skin and listens to that dismal noise.

Our camp at Fashur was very hot and dusty. In the afternoon of the next day we started for a march of twenty-two miles, which would take us half the way from the Atbara to Kassala. We turned our backs to the river and crossed a plain covered by dried grass and mimosa scrub, which is probably a swamp during the rainy season. Our tents were pitched on the open, level ground after an exhausting journey, at the halting-place ordinarily used by those who follow this track. There were no habitations, and we saw none on the road. During the night a detachment of the Arab Battalion passed our encampment, on its way to protect villages against slave-raiders.

On March 15 we arrived at the River Gash just outside Kassala. Its course here lies in a flat, sandy bed, which was quite dry. During the journey—eighteen and a half miles—we saw eight ostriches. The Gash, beyond Kassala, splits into numerous small streams which, even in the rainy season, sink into the ground and disappear completely. Sir Samuel Baker has given an account of the river at that time of the year, which shows its importance in relation to produce and water-supply in the district.

"As we approached within about twenty-five miles of Kassala, I remarked that the country on our left was in many places flooded; the Arabs, who had hitherto been encamped in this neighbourhood during the dry season were migrating to other localities in the neighbourhood of Soojalup and Gozerajup with their vast herds of camels and goats. As rain had not fallen in sufficient

KASSALA HILL AND MARKET-PLACE.

[*See p.* 205.

quantity to account for the flood, I was informed that it was due to the river Gash, or Mareb, which, flowing from Abyssinia, passed beneath the walls of Kassala, and then divided into innumerable ramifications; it was eventually lost, and disappeared in the porous soil, after having flooded a large extent of country. This cause accounted for the never-failing wells at Soojalup—doubtless a substratum of clay prevented the total escape of the water, which remained at a depth of forty feet from the surface. The large tract of country thus annually flooded by the river Gash is rendered extremely fruitful, and is the resort of both the Hadendowa and the Hallonga Arabs during the dry season, who cultivate large quantities of dhurra and other grain. Unfortunately, in these climates, fertility of soil is generally combined with unhealthiness, and the commencement of the rainy season is the signal for fevers and other maladies." *

We were now close to the curious and isolated eminence called Kassala Hill, which stands within the border-line of the colony of Eritrea. We had seen this towering landmark when we were more than forty miles distant from it, and a whole day's travelling had seemed to bring us no nearer to it. It offers a splendid sight in the desert, when the setting sun shines on the bare pile of red granite. I was told that in the days when the Italians occupied Kassala, Alpine climbers, with elaborate paraphernalia, had attempted the ascent, but failed in all cases to reach the summit. There is a superstition among the natives that any one who tries to scale the heights will die shortly afterwards, and this belief has been confirmed among them by the death of Colonel Collinson, a former Moudir of the town,

* "The Nile Tributaries of Abyssinia," p. 66.

who made the last endeavour of which I heard, and succumbed to fever six months afterwards.

We rose early next morning (March 16) and visited the British officers at the Moudirieh. Needless to say we were in touch with a civilized *cuisine* again, and nothing teaches a man to appreciate a good lunch in a cool room like a journey in the desert. In the afternoon I visited the hospital, which was in charge of Dr. Ensor. It is a most instructive place to a medical man; for here patients are gathered together from all parts of the Eastern Soudan. And it affords a proof of the benefits of Anglo-Egyptian administration that is beyond cavil. I saw cases of many interesting tropical diseases, about which the doctor and I had a long talk in the evening. We dined at the mess of the 11th Battalion of the Egyptian army—once more among our fellow-countrymen.

On the next day we visited Cristo's, the Whiteley's of Kassala, and made purchases for the remainder of the journey. At sundown I watched the 11th at tattoo. This regiment had a good band, composed of Soudanese blacks, and it gave one a lively pleasure to hear European music again, though I must admit that they played the Old Hundredth arranged as a march! In the evening the Governor had a dinner-party in honour of St. Patrick's-day, and a number of officers were present. The place might have been an Indian station instead of Kassala.

On March 18 my companions left the camp very early, accompanied by another English soldier who was visiting the town, to make an inspection of the bed of the Gash above Kassala. In the evening they rode out in the opposite direction to see the end of its course in the tract in which it is absorbed. Possibly the

THE MOUDIRIEH AT KASSALA. [*See p.* 205.

TENTS PITCHED IN THE ENCLOSURE OF THE MOUDIRIEH AT KASSALA.
[*See p.* 205.

subterranean waters, free from evaporation, will afford an invaluable reservoir for irrigation and the maintenance of stock in the future.

The Governor of Kassala had a small black servant, about thirteen years of age, called Fadl Mullah (Courtesy of the Prophet). The lad had been picked up in a deserted Dervish trench during the battle of the Atbara, where he was found tied to a dead camel and covered with blood. The Governor told us that he was "as sharp as a needle" and most zealous in service, never sleeping in the afternoon, and always running when he was sent upon an errand. He only gave trouble in one way, and this was that once in every two months or so he would come to the Moudir and declare that he had found his father. When the putative parent had disclaimed the bond, and the Governor had said that he would hear no more about it, the boy would acquiesce. But he was sure to discover another "father" a few weeks later.

While we were at Kassala an Italian officer arrived in the town from Eritrea. No one had received notice of his coming, and no one seemed to know what his business was. I sat next to him one night at dinner, and he told me that he was a lieutenant in the Mountain Artillery, and was using his leave to make a journey to the junction of the Atbara and Settite Rivers. He would return thence direct to his battery.

On the 20th we struck our camp and left Kassala in the afternoon, turning our faces towards the Atbara again. We had the pleasure of the company of the Governor and another English officer during the next two days' journeys, which were hot, tiring, and uneventful. We bade farewell to our friends on the afternoon of the 22nd. On the following day we marched in the direction of the Goz Regeb hill. This singular rise of ground in the desert appeared and disappeared as we moved down imperceptible slopes and then ascended again. We saw the mirage all around us. The soil in this region was shingly. At Goz Regeb there was a two-roomed rest-house. It was of baked mud, and was the most solid building that we had seen since we left Khartoum. We had been told at Kassala that it belonged to the Slavery Department, and that we might make use of it as it was then unoccupied. That night there was a high wind, and among its *ludibria* was Crawley's sponge, which, oddly enough, was found next day in the Atbara about half a mile from our halting-place.

On March 24 Dupuis climbed Goz Regeb Hill and took photographs of the curious balanced granite blocks which stand upon it. At a distance many of them look like figures of men, and at a nearer range like worn statues. But it is certain that they have not been placed in their position by human agency. I am unable to offer any conjecture as to their origin or geological relation to their surroundings.

An incursion of great numbers of Arabs from the south, with their flocks and herds, into the region around Goz Regeb takes place regularly at the commencement of the rainy season. They are then compelled to come to this district to avoid the seroot fly *—

* "This peculiar fly is about the size of a wasp, with an orange-coloured body, with black and white rings; the proboscis is terrific, it is double, and appears to be disproportioned, being two-thirds the length of the entire insect. When this fly attacks an animal, or man, it pierces the skin instantaneously, like the prick of a red-hot needle driven deep into the flesh. At the same time the insect exerts every muscle of its body by buzzing with its wings as it buries the instrument to its greatest depth. The blood starts from the wound immediately, and continues to flow for a considerable time; this is an attraction to other flies in great numbers, many of which would lay their eggs upon the wound." ("The Nile Tributaries of Abyssinia," pp. 184, 185.)

GOZ REGEB GRANITE STONE, MIMOSA SCRUB IN THE DISTANCE.

From a Photograph by Mr. C. E. Dupuis. [*See p.* 208.

GOZ REGEB STONES. [*See p.* 208.

From a Photograph by Mr. C. E. Dupuis.

the tsetse of the Soudan—which is fatal to all live stock except goats. Obviously, this necessary migration is an important fact both in relation to politics and campaigning in the Soudan. Sir S. Baker came in contact with the movement northward.

"The commencement of the rainy season was a warning to all the Arabs of this country, who were preparing for their annual migration to the sandy and firm desert on the west bank of the river, at Gozerajup; that region, so barren and desolate during the hot season, would shortly be covered with a delicate grass about eighteen inches high. At that favoured spot the rains fell with less violence, and it formed a nucleus for the general gathering of the people with their flocks.

"The burning sun, that for nine months had scorched the earth, was veiled by passing clouds; the cattle that had panted for water, and whose food was withered straw, were filled with juicy fodder; the camels that had subsisted upon the dried and leafless twigs and branches, now feasted upon the succulent tops of the mimosas. Throngs of women and children mounted upon camels, protected by the peculiar gaudy saddle hood, ornamented with cowrie shells, accompanied the march: thousands of sheep and goats, driven by Arab boys, were straggling in all directions; baggage-camels, heavily laden with the quaint household gods, blocked up the way; and fine bronzed figures of Arabs, with sword and shield, and white topes or plaids, guided their milk-white dromedaries through the confused throng with the usual placid dignity of their race, simply passing by with the usual greeting, '*Salaam Aleikum*' (Peace be with you).

"It was the Exodus; all were hurrying towards the promised land—the 'land flowing with milk and honey,' where men and beasts would be secure not only from the fevers of the south, but from that deadly enemy to camels and cattle, *the fly;* this terrible insect drove all before it." *

During the wet season the Mouderir (Government) of Kassala is transferred to new buildings which have recently been erected for its reception at the village of Goz Regeb, and there is a general movement of the Europeans and the wealthier native inhabitants of the town to the same place. The exercise of jurisdiction and the collection of revenue in this region are made difficult by the annual migration of which Sir Samuel Baker wrote. Great numbers of people from different parts of the country collect now, as in former days, within one district in a very short space of time, the comminglement of flocks and herds adds to the confusion, and it is no simple administrative task to deal with a shifting population of this magnitude. If the seroot fly were exterminated, as certain species of noxious African mosquitoes have been, this yearly exodus would no longer take place.

On March 25, we camped beside the river at a pleasant place well shaded by trees. I took my rod to a pool and landed a couple of fish, one about six pounds and the other twelve pounds in weight. They were of the same species as my fifty-pound prize, and both showed fight. They came into shallow water after plenty of coaxing, and I got them ashore without aid. My experience that evening shows that a diary is kept with difficulty in the Soudan. I sat on my bed to write the notes of the day, and held a candle aloft in my left hand to keep it out of reach of the draughts which blew the flame in all directions if I lowered it. With the right hand I used

* "The Nile Tributaries of Abyssinia," pp. 107, 109, 110.

ROCKS AT GOZ REGEB. [See p. 208.

From a Photograph by Mr. C. E. Dupuis.

ROCKS ON GOZ REGEB HILL. [See p. 208.

From a Photograph by Mr. C. E. Dupuis.

my pencil, and kept clearing away the insects that flew upon my face and swarmed about the light. They seemed to exercise no choice, but flew indifferently into one's mouth or eyes or into the flame, and they were nastiest when moribund, but still active, after resolutely passing through the fire.

On the 26th we halted, after marching from five in the morning till eleven, at a spot a few miles distant from the river, where there are three rest-houses—one for the chiefs of the party which is travelling, one for the servants, and one to serve as kitchen. Gazelles were plentiful in this district, but we were unable to use our rifles on account of the throng of Arabs, cattle, and sheep in the neighbourhood. It was interesting to watch the manner in which the goats feed at this season. They are tended by half-naked Soudanese boys who carry long crooked staves. With these they pull down the branches of the mimosas, and the goats browse the leaves and twigs. They also stand upon their hind legs, resting their forelegs upon boughs, and so reach the foliage, keeping the branches down with their hoofs, while they eat the succulent new shoots. When they are seen from a distance in this attitude they exactly resemble people. In this region the villages are very small—mere hamlets consisting of a few huts, and very little ground is under cultivation. The land is used for pasture throughout the district, and is apparently common to all comers.

On March 27, we travelled about thirteen miles and halted at a rest-house. We had reached a zone of lower temperature—the maximum at midday was 89°—and the journeys were no longer unpleasant. The river-side scenery is interesting and beautiful in the lower as well as in the higher reaches, and the dome-palms become more numerous as one advances northward and add to the charm of the banks. There was no other vegetation when we saw the country except low-grown mimosas and mimosa-scrub. Sand stretches away on either side from the course of the river, and we crossed few khors after leaving Goz Regeb, for the rainfall in this region is absorbed in the soil and is insufficient to produce torrents which would scour a course towards the stream.

In the afternoon I took my fishing-tackle and tried my fortune in a pool from which I landed two fish, one of about fifteen pounds and one of about five pounds, and a crocodile tried his luck upon the bank and nearly caught a man. Before I set out with my rod I had been warned by the sheikh of the village near which we had encamped that the beasts were very dangerous in this part of the Atbara, and I kept a good look-out for signs of them. It happened, however, that just as I had made a cast, Dupuis, who had been out with his rifle, passed behind me and sang out, "Have you caught anything?" I stepped back briskly, as good luck would have it, and answered, "Have *you* shot anything?" and at the instant the crocodile lashed at my legs with its tail to strike me into the water. It had been awaiting its chance to take me off my guard, and I had had no inkling of danger. I judged the length of the reptile to be about twelve feet. It certainly had a sound sporting plan and made a smart dash; for it only missed me by a few inches. The tails of these creatures are very powerful, and if the one that had been stalking me had judged distance a little better, I should certainly have been knocked into the pool.

A little later I heard in Cairo that a week after my adventure, a woman, who was filling a water-sack close

A MIRAGE, SHOWING GOZ REGEB HILL IN THE DISTANCE, ALSO NUMEROUS GAME TRACKS. [*See p.* 208.

to the place where I had stood, turned round to answer some men who were exchanging chaff with her; as she did so, a crocodile struck her on the hip with its tail, lashed her some distance into the water, and immediately dragged her under. I was told that they use the tail-trick only with human beings and always seize beasts by the snout. But this account does not quite agree with the observations of Sir S. Baker, who made a very careful study of the habits of the crocodiles in the rivers of the Soudan.

"The crocodile perceives, while it is floating on the surface in mid stream, or from the opposite side of the river, a woman filling her girba,* or an animal drinking, etc. Sinking immediately, it swims perhaps a hundred yards nearer, and again appearing for an instant upon the surface, it assures itself of the position of its prey by a stealthy look; once more it sinks, and reaches the exact spot above which the person or animal may be. Seeing distinctly through the water, it generally makes its fatal rush from beneath—sometimes seizing with its jaws, and at other times striking the object into the water with its tail, after which it is seized and carried off.

"The crocodile does not attempt to swallow a large prey at once, but generally carries it away, and keeps it for a considerable time in its jaws in some deep hole beneath a rock, or the root of a tree, where it eats it at leisure." †

The fish which I had caught were of the same species as those which I had previously landed, and proved very good at table in the evening. It was our custom to dress for dinner—in pyjamas.

On March 28 our journey lay through the desert,

* Water-sack. These are made from skins.

† "The Nile Tributaries of Abyssinia," pp. 240, 241.

and reminded us of our first marches towards Gedaref. The river banks with their dome palms were the only landmark. Elsewhere nothing was to be seen but sand, with patches of thinly-growing, very coarse grass or low mimosa scrub here and there, and mirage all around us.

Next day we travelled twenty-one miles to Adarama. On the way we passed at a distance a great zareba prepared by Osman Digna. This was protected by double walls, each about twelve feet high. There is an interval of some three feet between the two. They enclose a square space of sandy ground, and the sides have a length of about three hundred yards. The Dervish leader intended to occupy it after the battle of the Atbara. The defences are now slowly falling to pieces; for the walls are made of sun-dried mud, and as this becomes weather-worn and crumbles the wind carries it away, leaving gaps in the ramparts. The sight symbolizes much. Adarama was formerly one of the large towns of the Soudan, but now consists of a few mud huts. There is also a small garrison of about twenty soldiers. The place lies halfway between Gozerajup and Berber, and has a pleasant rest-house among the palm-trees close to the river. The bed of the Atbara is sandy here, and the average breadth is about four hundred yards.

On March 30 we reached Gumaza. I had the luck to bring down a fine buck gazelle while we were on the road. In the evening Dupuis and I made practice at the crocodiles in the river, and their numbers were shown by the commotion that they made in the pools. The next day we pushed on across the desert towards Berber and slept in the open. On April 1 our road lay through similar country. Sometimes a gazelle would scamper away in the distance, but often nothing was to be seen but sand and mirage. And then suddenly we were

LORD KITCHENER'S BRIDGE OVER THE ATBARA, NEAR BERBER. [*See p.* 214.

confronted by vestiges of civilization. When we were not far from the Atbara battle-field we saw the unmistakable traces of a traction engine. About eleven o'clock dome palms were in sight, and we reached a rest-house beside the river half an hour later. After lunch we heard the jerky puffing of a petrol motor, and rushed out to see what was approaching. It was a heavy engine dragging a car, and the tracks which we had discerned were accounted for. The Englishman in charge told us that he had carried a tombstone to the battlefield; it had been erected on an officer's grave. He added that he could travel six miles an hour in the desert, but that petrol was very expensive in Berber, where it cost ten shillings a gallon; so he had come to the rest-house to await a troop of forty mules, which were to drag the engine and car back to the town.

On April 2 we reached the bridge over the Atbara—*monumentum ære perennius*, if Lord Kitchener's energy and perseverance receive their deserts from posterity. It was strange to us to see a railway train, and hear the whistle of a locomotive once more. On the next day we arrived at Berber. Here we were in full contact with civilization, and here, practically, the duties of the expedition ended, for we followed the ordinary train and steamer route to our journey's end. Here, too, I will take leave of my reader, hoping that the long excursion from Khartoum has not wearied him, and will only add my conviction that a splendid future lies before the Eastern Soudan, and before Abyssinia, *after* the clouds have gathered over that country and burst.

A NOTE ON THE RELIGION, CUSTOMS, ETC., OF ABYSSINIA

IN *The Pall Mall Gazette* of August 19, 1904, an interesting account of an interview with M. Hugues Leroux was published. M. Leroux had just returned from a "diplomatic and scientific mission" to Abyssinia. He had been the bearer of an autograph letter from M. Loubet to the Emperor Menelek. M. Leroux brought back much interesting information which concerns only the present circumstances of the country—*e.g.* that the Emperor "is striving to follow out what he himself calls a policy of good-humour. He wants to be on good terms with all the world, but he does not want to be devoured. He is anxious to open up Abyssinia, but to commerce, not to conquest." And again, "he has at last granted permission for the railway from Djibouti, which hitherto went only as far as Harrar, to be carried up to Addis Abiba. He will probably refuse the British and Italian requests to build branch lines connecting this railway with Zeila and Massouah, the towns on the coast to the north and the south of Djibouti which belong respectively to Great Britain and Italy, his comment upon this scheme being, 'I would like one ladder up to my front door, but not three.'"

M. Leroux had been entrusted with an autograph reply from the Emperor Menelek to the President of the French Republic. But he bore away from Ethiopia a much more interesting document, which—if it is genuine and is rightly described by the traveller and the interviewer—is invaluable for the clue that it gives to the origin of the Abyssinian nation. It is said to be "a most extraordinary Abyssinian manuscript, which is certain to make a great sensation when the translation of it is published, for it throws a flood of light upon the early history of the Jews. The Abyssinian monarchs have always considered this manuscript to be among their most sacred possessions. It was appropriated by the English at the capture of Magdala, but on Emperor John succeeding to the throne, he opened diplomatic negotiations and recovered it from the British Museum, where it had been deposited. It is written in the Geez language, which is of Hebraic origin, of great antiquity, and known only to the priests. It relates that after the plagues of Egypt some hundreds of thousands of Jews did not follow Moses across the Red Sea, but went west, down the Blue Nile, to found a kingdom of their own, which they called Saba—'Saba' merely meaning west. These were the original Abyssinians."

We are told that "the original of this manuscript had remained seventeen hundred years in an island of Lake Zouai, whose inhabitants are still of the purest Jewish type, but they were Christianized at a very early date, and their religion has now degenerated into fetishism. They knew nothing as to the contents of the manuscript, which they nevertheless regarded as a very precious object." Whatever may be the real antiquity of this document its existence and the importance attributed to it seem to give evidence of a tradition of great age, and this tradition offers a credible explanation of the fact that such a people as the

Abyssinian is found in the mountains of East Africa. No doubt the Habashes are now, almost universally, a race of mixed blood, but they possess a national character which is quite distant from that of the neighbouring tribes. The Semitic invasion may have taken place immediately after the captivity of the Jews in Egypt or at an earlier or later date, but it was powerful enough to affect the whole polity and population of Ethiopia.

Generally, writers upon Abyssinia ascribe the incursion to the reign of King Solomon, and this is in accordance with a legend that is believed throughout the country. The Queen of Sheba was, of course, the Queen of Ethiopia; her visit to the wise and much-married king provided her with an heir; "her son Menelek was in due time transmitted to his august sire. The young prince was duly instructed in all the mysteries of Jewish law and science, and being anointed king under the name of David, he was returned to his native land, escorted by a large suite of the nobles of Israel, and a band of her most learned elders under the direction of Ascarias, the son of Zadok, the High Priest.

"The gates of the temple of Jerusalem were left unguarded, and the doors miraculously opened in order that the holy ark of Zion and the tables of the law might, without difficulty, be stolen and carried away. The journey was prosperously performed, and the Queen-Mother, on resigning the reins of authority to her son at her death, about nine hundred and seventy years before the birth of Christ, caused a solemn obligation to be sworn by all that henceforth no female should hold sway in the land; and that those princes of the blood royal upon whom the crown did not devolve should, until the succession opened to them or during the natural term of existence, be kept close prisoners on a lofty mountain; a cruel and despotic enactment which, through a long succession of ages, was jealously observed." *

"The impetuous zeal of the emigrants," wrote Mr. Stern, "found ample scope for its loftiest inspiration in the new world to which they were transplanted, and in the course of a few years the worship of the God of Israel extensively supplanted the idolatries of Ethiopia.

"From these vague traditions," the same writer prudently added, "in which truth and fiction are inextricably jumbled together, the inquirer does not gain much trustworthy information on the history of Ethiopia and the settlement of the Jews in that country." †

Mr. Augustus Wylde visited a charming and sequestered spot where, according to one legend, the stolen ark was deposited. "We halted for lunch just *vis-à-vis* to the first sources of the Taccazze. . . . Just before reaching the sources, on a hill on the north side of the valley, is the church of Chevenan Gorgis, in a splendid grove of juniper trees; immediately above the sources on the hill is another church, also surrounded by juniper trees, dedicated to Debbessa Jesu; tradition has it that when Menelek, the son of the Queen of Sheba by King Solomon, came from Jerusalem with the ark, it was placed on the ground at this spot, where he camped, and these springs gushed forth, and he immediately ordered a temple to be built on the spot.

"On leaving this camp Menelek commenced his march towards the east, and on the bearers of the ark putting it down, after about an hour's march, they found that they could not move it, as it firmly adhered to

* "The Highlands of Æthiopia," by Major W. Cornwallis Harris, of the H.E.I.C.'s Engineers, 1844, vol. iii. pp. 3, 4.

† "Wanderings among the Falashes," p. 185.

the ground. This spot is supposed by some of the Abyssinians to be the true resting-place of the ark that was brought from Jerusalem; there is a church built over the spot called Eyela Kudus Michael. . . . It is nearly impossible for a stranger to obtain admittance to this church, and the place in the Holy of Holies where the ark is supposed to rest is shown to no one. This ark cannot be in two places; the people of the north declare it is in the sacred grove of Axum in the church of Selata Musser (place of Moses), and the priests of the Eyela district declare it is in their church, so they always quarrel and wrangle over this vexed question, the same as European priests do over their sacred relics." *

Perhaps the soundest opinion which can be formed under present conditions as to the origin of the Abyssinian people has been expressed by Mr. Wylde :—" The race, as the name Habash or Abyssinian denotes, is a mixture, undoubtedly of very long standing, but most likely of Jew with the inhabitants of Southern Arabia and the non-negro races of Eastern Africa." †

According to tradition, Christianity was introduced into the country in the fourth century.

" In the year 330 after the birth of our Saviour, Meropius, a merchant of Tyre, having undertaken a commercial voyage to India, landed on the coast of Ethiopia, where he was murdered by the barbarians, and his two sons, Frumentius and Edesius, both devout men, being made prisoners, were carried as slaves before the Emperor. The abilities, the information, and the peaceable demeanour of the brothers soon gained not only their release but high office in the court; and living in the full confidence of the monarch until his decease, and subsequently under the protection of the Queen-

* "Modern Abyssinia," pp. 352, 353. † Ib. p. 16.

mother, the good-will of the entire nation quickly succeeded. The work of conversion was commenced, and proceeding with wonderful rapidity and success, a thriving branch was shortly added to the great Eastern Church.

"Bearing the happy tidings, Frumentius appeared in Alexandria, and was received with open arms by the Patriarch Athanasius. Loaded with honours, and consecrated the first bishop of Ethiopia, a relation was thus happily commenced with Egypt, which has remained firm and friendly to the present day, and throughout fifteen centuries has bestowed upon a Coptish priest the high office of Patriarch Abouna (*i.e.* our Father) of the Ethiopic Church.

"On his return to the country of his hopes, Frumentius found that the spark of life had spread rapidly throughout the gloomy darkness of the land. Baptism was instituted, deacons and presbyters appointed, churches erected, and a firm foundation laid whereon to establish the Christian religion in Abyssinia.

"During the succeeding century, priests and apostles, men of wonderful sanctity, flocked into the Empire from all parts of the East, and miracles the most stupendous are related in the legends of those days. Mountains were removed, and the storms of the angry ocean stilled by the mere application of the staff. The adder and the basilisk glided harmless under foot, and rivers stayed their roaring torrent, that the sandal of the holy man should remain unstained by the flood. Aragáwi raised the dead—the fingers of Likános flamed like tapers of fire—Samuel rode upon his lion; and thus the kingdom of Arwé, the old serpent of Ethiop, was utterly overthrown." *

The conservatism of the Abyssinian people in matters

* "The Highlands of Æthiopia," vol. iii. pp. 84–86.

of religion is well illustrated by the fact that throughout the many and violent vicissitudes of their history since the introduction of Christianity they have always received their Patriarch from the mother-church. His position is one of great authority. "The chief man in the country after the king of kings is the Abouna, or archbishop, the head of the Church; without the Abouna no king can be crowned, and it is he that, at his own or the king's wish, can excommunicate any of his subjects, or the king himself, if necessary, and then the king can only rule by the strength of his followers who adhere to him. These archbishops come from the Coptic Monastery at Alexandria or Cairo, and when they reach Abyssinia, they never leave it on any consideration." * Moreover, "the Abyssinian Church, in common with all other Christian communities in Asia and Africa, is strictly episcopal. The Abouna, or primate, who is consecrated to his office by the Patriarch of Alexandria, the revered successor of St. Mark, can alone confer the priestly title. Every candidate, before presenting himself for ordination, must have acquired some knowledge in the reading of the sacred language of Ethiopia, and in the complicated ceremonies of the liturgical service. On the day appointed for ordination the primate, in full canonicals and seated on the episcopal throne, receives the applicants for the sacred office. All being properly ranged before the chair of St. Mark, each candidate, instead of the imposition of hands, receives the Abouna's consecrating breath. Former archbishops, less scrupulous than the present successor of Frumentius, indiscriminately breathed on all, whether qualified or not, who could pay the requisite fee of two salts—fourpence.†

* "Modern Abyssinia," p. 157.

† Mansfield Parkyns wrote: "The ordination of priests and deacons

"Deacons are selected from among boys, who are only allowed to serve in the Church till they attain the age of twelve or thirteen; after that period their purity of life is suspected, and they are no longer considered fit to approach the sacred shrine of the *tabot*.* The bishop and monks may not marry, while the priests may; and as, on the death of their wives, they cannot contract a second alliance, the reverend wooers invariably choose for their partners the most robust and sprightly lasses in the land.

"The *debterahs*, or scribes, constitute the lowest but the most influential body in the Church. These worthies enjoy no ecclesiastical rank and are under no ecclesiastical discipline, and yet no service can be properly performed unless they take part in it. Their chief duty consists in chanting the Psalms and Liturgy, but their uncouth gesticulation and discordant shouting, instead of elevating devotion, tend rather, at least in European estimation, to convert the service of God into a sinful burlesque and the sanctuary into a bedlam. The scanty learning of the country is exclusively monopolized by this order; and they are so proud of their erudition that they deem it a disgrace to exchange, by the breathing of the Abouna, the proud title of *debterah* for the less learned appellation of *kas*, or priest.

is, I believe, tolerably simple; for instance, I have been told that, on the arrival of the present Abouna from Egypt, the candidates, who are only required to be able to read a little, were collected in a mass near the place where he was. The bishop then went through some ceremony, and ended by pronouncing a blessing, and blowing in the direction of the assembled crowd, who were thus all ordained. Among these was a woman with her child in her arms, who had come thither from motives of curiosity. She, too, was of course ordained; but I don't remember hearing that she ever officiated. If a priest be married previous to his ordination, he is allowed to remain so; but no one can marry after having entered the priesthood." ("Life in Abyssinia," p. 294.)

* See p. 240.

"These literati, notwithstanding their better acquaintance with the sacred volume and the lives of the saints, are considered the most arrant scoundrels in the land. Gondar, which contains a considerable number of the fraternity, is notorious for the dissolute profligacy of its inhabitants; and it is proverbial throughout the country that wherever *debterahs* abound there vice and immorality thrive." *

Mansfield Parkyns found that many of the *debterahs*, or *defterers*, professed to practise witchcraft. He wrote: "There are persons in the country who are supposed to be able to call up spirits and obtain from them any information they may require. These men are mostly *defterers* (or Scribes), and are regarded with a certain amount of rather dubious respect. The scene of their incantation is generally some ravine, with a stream running through it. As I have only heard the reports of the common people on the subject, I cannot describe accurately what ceremonies are observed, or what form the devil assumes on quitting the water at their summons. My informants, however, assured me that he came as a great chief, with the usual train of shield-bearers behind him and gunners before him." †

The Abyssinian liturgy is not impressive, according to European ideas. Stern was present at one of the services, and gave a lively description of it. "The deafening tom-tom of the *negareet*, intermingled with the nasal chorus of a host of *debterahs* in varying cadences, reverberated on my ear. This indication that the service had already begun put a stop to our contemplation of the grand mountain scene, and in a most reverent mood we hastened to the uproarious sanctuary.

* "Wanderings among the Falashas," pp. 307, 308.

† "Life in Abyssinia," p. 321.

A wooden gate in a circular wall brought us within an open grassy space that formed the cemetery, and the spot where, adjoining to the church, rises the Bethlehem in which the priests prepare the eucharistic bread. A crowd of men and women, as if in some place of amusement and dissipation, were spread in picturesque knots over the soft turf. The majority were busily plying their tongues, but I fear from the burst of merriment which now and then broke from one or the other of those animated groups that their conversations had very little to do either with religion or the service in honour of the saint. Not belonging to this impure class, who are justly excluded from the interior of the sacred edifice (*i.e.* those whose marriage is secular and dissoluble), we mounted a few steps, and then, through a partition occupied by the laity, stepped into a second compartment concentric with the outer one, and there found ourselves at the porch of the enclosure which constitutes the *sanctum sanctorum*.

"Crowds of priests and *debterahs* thronged the whole of that corridor. The *debterahs* constitute the choir in all their churches; and their devotionless mien, as they chanted to the monotonous sound of the *negareet*, quite excited my indignation. During the maddening noise created by the *debterahs*, the priests, robed in gaudy canonicals, were exerting to the utmost their cracked voices in intoning the liturgy and psalms. At certain intervals an ecclesiastic, clad in his garish finery, and attended by an incense-waving boy-deacon and a bearer of the Ethiopic Gospel, marched out of his sanctum into the cemetery to edify the godless and profane multitude by reading to them a portion of Scripture in an unknown tongue. A reverend gentleman near me, who was evidently a great Church dignitary, as he wore a huge turban, and had a very stupid unmeaning face, pointed his gaunt fingers towards the daubed walls, and condescendingly inquired whether I knew St. George and Miriam, the mother of God? . . .

"Unable to perform any religious rite without the savage accompaniment of tinkling keys and other discordant sounds, the sacrament of the mass (*Corban*), the most solemn service of the Church, is also performed amidst the most confused and distracting clangour. The liturgy and consecration service over, all, except the communicants, leave the place of worship. These now approach the vestibule of the holy of holies, where the officiating priests, enveloped in clouds of incense, are busily occupied in washing their hands—this ceremony is copied from Pilate's example. The water, which this act sanctifies, must not be spilled on the ground, but, as a regeneration emblem, it is sprinkled on the bended heads and the garments of the faithful, whilst the priest says, 'If you think that I have now cleansed your garments and purified your bodies, and yet continue to cherish hatred and malice in your hearts, I tell you that the body of Christ will prove to be a burning fire to consume you, and His blood a bottomless sea to drown you!' After this exhortation the *tabot*, the substitute for the altar, is taken out of the holy of holies, and each communicant receives a small piece of wheaten bread and a spoonful of raisin wine. To prevent the desecration of the sacred elements every one, before he quits the church, drinks a cup of water, and also refrains from expectorating that day." *

The ecclesiastical art of Abyssinia has already been mentioned. Stern has given an account of a singular instance of its development and of an equally strange

* "Wanderings among the Falashas," p. 214 *seq.*

result of the devotional use of pictures. He visited "Kudus Yohannes, which next to that at Quosquam, is the handsomest and most gorgeously bedaubed of the forty-four churches in and around Gondar." Here "one aspiring artist, weary perhaps of the antiquated cherubims and saints, the blazing flames and leafy bowers, which are the ordinary ornaments of the churches, had sought immortal fame by painting quite a new subject—the Migration of the Israelites. In his picture he represented them marching in soldier-like attitude over the heaving and surging waves of the Red Sea, clad in British uniform, with muskets and bayonets on their shoulders." Again, Major Harris, at the time of his mission to Sahela Selassie, who was then Negus, had presented His Majesty with a portrait of Queen Victoria. This "lay securely guarded amidst the regalia of the Ethiopian Empire till the death of the despot. His heir and successor, Hailu Malakot, to atone for some indiscretion, presented it in pious contrition to the Cathedral Church, and there the people flock on all grand festivities to worship it as the representation of the Virgin Mary." *

Reference has been made to certain Abyssinian saints, to whose exploits Major Harris alluded. But these are neither the most notable nor the most venerated in the ecclesiastical chronicles. Tekla Haimanot—*facile*

* "Wanderings among the Falashas," pp. 201, 202. Major Harris inspected the Cathedral of St. Michael at Ankóber, and in his description of it wrote, "In the holy of holies is deposited the sacred *tabot*, consecrated at Gondar by the delegate of the Coptic patriarch; and around the veil that fell before this mysterious emblem there hung in triumph four sporting pictures from the pencil of Alken, which had been presented to the king. They represented the great Leicestershire steeple-chase, and Dick Christian, with his head in a ditch, occupied by far the most prominent niche in the Cathedral of St. Michael!"

princeps—"is said to have converted the devil, and induced him to become a monk for forty days, though what became of him afterwards we are at a loss to know. The same holy man, wishing to ascend a steep mountain with perpendicular sides was accommodated in answer to prayer with a boa-constrictor, which took him up on its back." * Before the birth of this saint, "a glorious light rested for several days over the parental house. At the baptism of the child the priest was so dazzled by its supernatural beauty that, lost in admiration, he dropped the babe, and might have killed it, had not an invisible hand kept it suspended above the hard floor." In due course Tekla Haimanot became a monk, and "the mortifications, self-imposed penances and incredibly long fasts which followed his initiation into the monastic brotherhood are faithfully recorded in the annals of the Church for the edification of the faithful." Later, he visited the Holy Sepulchre, after marvellous adventures among the Moslems, and was subsequently ordained priest by the Copt Patriarch in Egypt. He then proceeded to the Galla country, where his mother had been kept prisoner, and converted the heathen by hundreds of thousands. On one occasion he repaired to the monastery of Debra Damo, but here "the brotherhood did not much sympathize in the general jubilee that greeted the austere monk. The monastery stands on the summit of a perpendicular rock, and, being quite inaccessible, no visitor can reach it unless drawn up by a rope. The Evil One did not like the ascetic to tamper with the merry fellows on the rock, and, to effect his wicked purpose, he maliciously cut the frail support when Tekla Haimanot was in mid-air, and probably he would have been dashed to pieces in a ravine below, had not immediately six wings unfurled themselves under

* Dufton, "A Journey through Abyssinia," p. 90.

his garb and borne him aloft. In commemoration of this miraculous volant power, the saint is represented, in most of the churches dedicated to him, as nearly smothered in a profusion of gorgeous plumage."

Tekla Haimanot's ascetic practices in "tenantless wastes and malarious jungles" did not suffice him as the years advanced. "He hit upon an original idea of mortifying the flesh. There is in Shoa a small lake, which the saint in his peregrinations had often passed. To these waters he now repaired. The good people who followed him from all parts to hear his discourses and to obtain his blessing entreated him not to expose his precious person to the alligators and other aquatic monsters; but the holy man, who knew that all his exploits for the glory of the Church had not yet been accomplished, fearlessly stepped into the deep. Seven successive years he continued in the water, and probably he would have expired on his liquid couch, had not one of his legs dropped off. The clamour for this valuable relic created quite a dissension in the Church, but the monarch judiciously put a stop to the fierce war between the rival claimants by ordering it to be kept as a Palladium in the royal metropolis. This sacred talisman possesses more wonderful sanitary virtues than all the drugs in the universe. Patients from every province of the country visit the shrine to make votive offerings and to quaff the healing water in which the saint's leg is weekly washed." *

That the miracles above related are stock miracles —if one may use such a phrase—and that the same legends are connected with more than one saint is evident from the following *gesta sanctorum* cited by Mansfield Parkyns :—

* "Wanderings among the Falashas," p. 233 *seq.*

"Gabro Menfus Kouddos (Slave of the Holy Ghost) was a great saint from his birth; nay, more—he was born a saint. No sooner did he enter the world than he stood up, and three days after his birth he bowed his head thrice, saying in a distinct voice, 'Glory be to the Father, and to the Son, and to the Holy Ghost.' He never tasted of his mother's milk, nor during the whole period of his life partook of food or drink of any sort. Once, when praying on a mountain, he fell over a precipice two hundred cubits deep. Two angels immediately joined their spread wings under him to support him; but he refused their assistance, saying that he trusted to God alone for help. Another time he was ascending a very high mountain, and, being fatigued, the Holy Trinity came and assisted him. Many other equally absurd and almost blasphemous stories are related of him; at last, after a very long life, I forget of how many years, the Almighty sent Azrael, the angel of death, to take him. But the saint refused to die, saying, that as he had neither eaten nor drunk, he could not die. So all the saints came to him in turn for the purpose of persuading him to leave earth for Paradise. St. John the Baptist first addressed him, saying that he had gone the way of all flesh, notwithstanding his many privations and sufferings. Gabro Menfus Kouddos, however, at once met him with the answer, 'Yes; but *you* could not fast even for forty days, but fed on locusts and wild honey.' Thus he replied to all the saints, and at last even to the Virgin and our Saviour. Still, however, the decrees of the Omnipotent must be obeyed, and his life was taken from him; but then there was a dispute among the elements as to what was to become of his body. The earth refused to receive it, as he had never partaken of her produce. A similar refusal was made by the water, for he had never taken a drop within his lips. The fire had also equally strong objections. So the saint was restored to life, and taken up alive into heaven. His tomb is, however, shown at Zoukwahla in Shoa; but it is said to contain only one of his ribs, which, at the time of his ascent to heaven, he took out and left on earth as a memento for his followers.

"Abouna Aragawy was one of the nine missionaries sent to Abyssinia by St. Athanasius. His doctrine and the miracles which he wrought gained for him many followers; but from some of the unbelieving he suffered persecution. This is the account given by some historians; while others assert that, overcome by his popularity, he sought retirement from this world, in order to devote the remainder of his days to religious duty. Be it as it may, he came to the rock on which is now the celebrated monastery of Debra Damo. After walking several times round it, without finding any means of access to its summit, he prayed to the Almighty, who sent him an enormous boa-constrictor, which offered to carry him up in its mouth; but he said, 'I fear your mouth; turn round and let me take your tail.' So the snake did as he was desired; and the saint, holding fast by its tail, was drawn up to the summit of the rock in perfect safety. The snake having performed its duty offered to leave the saint if he wished it; but Aragawy begged that it would remain, making, however, the condition of its not alarming or destroying any of his disciples who might come to visit him. They then took possession of the caves and holes which are in the mountain, where they are by many supposed to be still living. Some, however, pretend that the snake is dead; but no one is so wanting in faith as for a moment to deny that the saint yet lives there, and will continue to live till the day of judgment. No curious person, however, dares to venture into the cave. The monks will not allow lights to be taken in; and the people assert that a spirit which protects the place will not permit any one who enters to come out alive.

"When I first went to Rohabaita it was reported that an immense snake had some years before been killed there by a hunter. The man was severely reproached by the priests, who said that the snake was the guardian angel of the place. It was reported to have been twenty-seven cubits, or forty feet in length, and was probably of the boa tribe." *

Another saint of remarkable powers was Tederos (St. Theodore). He resided at the hamlet near Rohabaita which now bears his name. Here, "on the face of the rock is the little hole wherein he lived; it is barely high enough for a person to squat in; and the marks worn in the stone by the crown of his head, the soles of his feet, and his elbows are still shown. His miracles were many. Among others, a leopard ate up his son, the saint, returning home, missed him, and set out in search of him. When in the forest he called aloud to him three times, and at the third time the leopard appeared: on seeing him, the saint guessed how the matter stood with the unfortunate youth; but, nothing discouraged, he coolly ordered the beast to return him safe and sound. Now this was rather difficult, as the leopard had, no doubt, half digested him; nevertheless, so great was the saint's power that the boy left the leopard's maw none the worse, perhaps rather the better, for having been dismembered and reconstructed. Being asked where a church should be built, the saint threw his staff, desiring the inquirers to build where they found the staff had

* "Life in Abyssinia," p. 297 *seq.*

fallen. After many days it was found several miles off in Serawi, on the other side of the valley, exactly on the spot where now stands the church of Debra Mariam.

"Many other wonderful stories were told me of his feats, but I have forgotten them. The most useful act attributed to him was that he caused the rock below him to become hollow, in order to receive the rain-water. The hollow still exists, though I should strongly suspect it to be of Nature's construction; or, if the saint had a hand in its design, he must have been a clumsy fellow, for with half the labour he might have made a place capable of containing twice the quantity of water." *

Parkyns was not sure of the parentage of the youth who was disgorged by the leopard. "I hope I am right in giving St. Theodore a son," he wrote. "I don't know whether his being a father would preclude his being a saint: however, it was a youth nearly related to him, if I remember the story right."

Some chroniclers "pretend that St. Matthew and St. Bartholomew actually visited the country. Some even go so far as to assert that the Virgin Mary herself, with the child Jesus, came into Abyssinia when she fled to Egypt, and show a place in a high mountain which is called her throne or seat." † Bruce wrote that "in the Synaxar, or history of their saints, one is said to have thrown the devil over a high mountain; another persuaded him to live as a monk for forty years; another had a holy longing for partridges, upon which a brace perched on his plate—martyrs ready roasted." ‡

"In many cases," Parkyns wrote, "the patron saint

* "Life in Abyssinia," pp. 160, 161.
† Ib. p. 277.
‡ "Life of Bruce," by Major Head, p. 266.

is preferred to the Almighty; and a man who would not hesitate to invoke the name of his Maker in witness to a falsehood would have difficulty in disguising his perjury if he were appealed to in the name of St. Michael or St. George. It is also a well-known fact, and most common of occurrence, that a favour besought in the name of God would often be refused, while, if the request were immediately after repeated in the name of the Virgin or of some favourite saint, it would probably be granted. This may be observed in the appeal of the common street beggar in Tigre, whose ordinary cry is, 'Silla Izgyheyr! Silla Medhainy Allam!' (For the sake of God! For the sake of the Saviour!)—while, if he be very importunate, he will change his usual whining tone, and add with persuasive emphasis, 'Silla Mariam! Silla Abouna Tekla Haimanout!'" (For the sake of Mary! For the sake of Tekla Haimanout!)

It has already been said that Tekla Haimanot and other early celebrities of the Church were monks. Monasteries are plentiful in the land, and there are also hermits who practise asceticism in desolate and lonely places. Mr. Herbert Vivian gained admission to a conventual house, and has given a lively description of it.* The sanctity of the Abyssinian monk or anchorite is in proportion to his prowess in mortifying the flesh, and this class of men still retains much religious *prestige* and influence in the land. But probably the pristine rigour of their life has been much relaxed, and the remarks of Major Cornwallis Harris are applicable at the present time.

"Education was in former days to be obtained alone from the inmate of the monastic abode, and a life of scanty food, austerity, and severe fasting was embraced

* "Abyssinia," p. 276 *seq.*

only by the more enthusiastic. But the skin-cloak and the dirty headdress now envelope the listless monk, who, satisfied with a dreamy and indolent existence, basks during the day on the grassy banks of the sparkling rivulet, and prefers a bare sufficiency of coarse fare to the sweeter morsel earned by the sweat of the brow." *

The same writer's account of the consecration of a friar is interesting. "The monk is admitted to the order of his choice by any officiating priest. A prayer is repeated, the skull-cap blessed with the sign of the cross, and the ceremony is complete. But a more imposing rite attends the oath of celibacy before the Abouna. The clergy assemble in numbers, and fires are lighted around the person of the candidate. His loins are bound about with the leathern girdle of Saint John, and the prayer and the requiem for the dead rise pealing from the circle. The *glaswa*—a narrow strip of black cloth adorned with coloured crosses—is then placed on the shaven crown, and shrouded from view by the enveloping shawl; and the archbishop, clad in his robes of state, having repeated the concluding prayer and blessing, signs with his own hand the emblem of faith over the various parts of the body." †

Consul Plowden wrote in a somewhat censorious mood of the Abyssinian regulars. "Monasteries are not wanting to complete the resemblance to the Roman Catholic Church and to the Middle Ages, where every immorality is practised; nor solitary hermits, who dwell in gloomy forests, feeding on roots, and exposed to ferocious animals, and who are sometimes as sincere as they are useless." He added, "nunneries alone are absent from the picture; though vows of celibacy are

* "The Highlands of Ethiopia," vol. iii. p. 139.

† Ib. p. 138.

sometimes taken, if rarely kept save at an advanced age."

With regard to religious sisterhoods he was in error. Bruce wrote that, "all women who choose to renounce acquaintance with men are allowed to turn priests, they then wear a skull-cap, like the men; and these priests, male and female, all pretend to possess charms of a nature both offensive and defensive, which are most generally believed in." * The same writer asserted that he and his party on one occasion, after passing the River Taccazze, "at last reached a plain filled with flowering shrubs, roses, jessamines, etc., and animated by a number of people passing to and fro. Several of these were monks and nuns from Waldubba, in pairs, two and two together. The women, who are both young and stout, were carrying large burdens of provisions on their shoulders, which showed that they did not entirely subsist upon the herbs of Waldubba. The monks, their *compagnons de voyage*, had sallow faces, yellow cowls, and yellow gowns." †

Bruce may have been mistaken as to the character of these young women; but not all who take the vows are strict in the observance of them. Mr. Augustus Wylde met a nun in the flesh, a lady of high rank in the country, whose charms were neither "offensive nor defensive." He wrote, "before getting to Mokareet we were joined by Ras Mangesha's young sister. She is supposed to be a nun and devote her time to charity and good actions, and she has asked permission of Hailou to join us and camp with us for the next three or four days. I now quote from my diary—

* Major Head, "Life of Bruce," p. 265.

† Ib. p. 268.

"I believe this is a got-up job, as Hailou had his hair replaited this morning and put on clean cloths, and the pair are evidently very old friends. She is dressed in the most spotless white garments of the very best quality, and is most clean and neat in her appearance. From under her clean white hood a pretty face appears with a pair of roguish merry eyes. . . . She sent over her pretty waiting-maid as soon as camp was pitched, asking me to come and pay her a visit, but I was too seedy with fever, so sent Schimper to make my excuses; when he came back his first remark in his precise manner of talking was: 'Oh, Mr. Wylde, I do not think she is a very good lady, she has asked me for many things, and she wants your umbrella.' Of course I had to send it, and next morning on the way to Merta I had some of the sweetest of smiles, and we kept up a long conversation; she had never seen an Englishman before, and she wanted to know if they all had red heads and moustaches and were big men . . .

"I found out that she was considered very 'rapid' even for Abyssinia, where the young ladies are fast, and that her brother had sent her to a nunnery as he could not keep her in order, and she absolutely refused to marry the men of his choice; she was changing her quarters from one nunnery to another, and was evidently sweet on Hailou."

But Mr. Wylde had met other nuns, "good pious old ladies at Abbi-Addi and Macalle, who fasted for over half the number of days in the year, and were perpetually praying and singing from the early grey dawn till late at night, and they seemed to live a calm and peaceful life in their beautiful natural surroundings, and bothered themselves very little with the troubles of this life, and passed their days in eating, sleeping, and praying, and doing what little acts of charity they could in the way of tending the sick and feeding any poor beggars that came to their houses." *

Many rites and ceremonies of Hebrew origin have been incorporated in Abyssinian Christianity, *e.g.* circumcision is practised. Considering the history and traditions of the people, this respect for Judaic ordinances is not surprising. Major Harris devoted a chapter of his book to "Abyssinian Rites and Practices which would appear to have been borrowed from the Hebrews," and most of the authors who have written about the country have alluded to the same subject. Perhaps the most remarkable example of the influence of Jewish doctrines is the veneration in which the *tabot* is held.

"The Jewish temple consisted of three distinct divisions—the forecourt, the holy, and the holy of holies. To the first laymen were admitted, to the second only the priest, and to the third the high priest alone. All entrance was denied to the Pagan, a custom which is rigorously enforced in Abyssinia; and her churches are in like manner divided into three parts.

"Eight feet in breadth, the first compartment stretches, after the fashion of a corridor, entirely around the building. It is styled *Kene Máhelet,* and, strewed throughout with green rushes, forms the scene of morning worship. To the right of the entrance is the seat of honour for priests and erudite scribes; and beyond this court, save on certain occasions, the bare foot of the unlearned layman cannot pass.

"*Makdas* is the second compartment. This is the sanctuary in which the priests officiate, and a corner is set apart for laymen during the administration of the holy supper, whilst a cloth screens the mysteries of the interior. Here also hang, arranged around the walls,

* "Modern Abyssinia," pp. 361, 362.

the bones of many deceased worthies, which have been carefully gathered from the newly opened sepulchre, and are deposited by the hand of the priest in cotton bags. By the nearest relative the first opportunity is embraced of transporting these mouldering emblems of mortality to the sacred resting-place of Debra Libanos, where the living and the dead are alike blessed with a rich treasure of righteousness, since the remains of Tekla Haimanot, the patron saint of Abyssinia, still shed a bright halo over the scene of his miracles upon earth.

"To *Kedis Kedisen*, the holy of holies, none but the Alaka * is admitted. Behind its veil the sacrament is consecrated, the communion vessels are deposited, and the tremendous mysteries of the *tabot*, or ark of the covenant, are shrouded from the eyes of the uninitiated. The gold of the foreigner has penetrated the secret of the contents of this box, which are nothing more than a scroll of parchment, on which is inscribed the name of the patron saint of the church; but the priest who dared to open his lips on the subject to one of his own countrymen would incur the heavy penalties due to the sacrilege." †

Another foreigner whose gold "penetrated the secret" of the *Kedis Kedisen* was Mr. Herbert Vivian. In company with a friend he was, for a consideration of two dollars, admitted to the *sanctum sanctorum* of the Church of St. Raguel at Entotto. "The custodians seemed to know very well that they were doing wrong, and displayed huge anxiety to get the thing over as quickly as possible. They hurried us up the rickety steps, opened the door of the sanctuary, took great care to keep us as

* The Alaka is the superior of a monastery or other head of an ecclesiastical establishment.

† Major Harris, "The Highlands of Ethiopia," vol. iii. pp. 134, 135.

far off as possible, and kept exclaiming, 'Now that is all. I hope you are satisfied. Let us go away again. You ought really to give us some more dollars for what we have done.' There was, however, really next to nothing to see. In the darkness I could just make out the tabernacle inside the holy of holies. It was a kind of ark, covered with cheap draperies and surmounted by a crucifix. A few lanterns lay about. That was all. At any rate, that was all that we could induce the priest to show us." *

Mr. Hormuzd Rassam has given a sketch of a *tabot* in his book, with the following description: "The *tabot* is a square block of wood on which the emblem of the cross, with some appropriate passage of Scripture, is sometimes represented. No church can be consecrated without it, and in the church from which it has been removed the Lord's Supper cannot be celebrated. It is placed upon the altar in the centre of the holy of holies, and by the ignorant clergy is regarded with as much veneration as the consecrated elements." † Annually, on the eve of the festival which commemorates Christ's baptism, the *tabot* is carried out of the church by the chief officiating priest, "and left all night covered in a tent or hut erected for the purpose near a stream or pool. At daybreak the following morning it is taken to the water, and after being sprinkled with a few drops is again enveloped in the church-cloth, and replaced under the tent until the time arrives for its formal restoration to the church—a grand religious ceremony, accompanied with singing of psalms and dancing." In this recessional solemnity, which Mr. Rassam saw, "the priests led the way, bearing various sacred symbols, and

* "Abyssinia," p. 283.

† "British Mission to Abyssinia," pp. 225, 226.

chanting psalms all the while, followed by a train of about three hundred men and a score of damsels, all dancing, singing, and screeching as they went." *

The appearance of Abyssinian churches has already been alluded to, but the following general description, given by Major Harris, may serve to bring the principles of their construction clearly before the reader. "There are, perhaps, more churches in Abyssinia than in any other part of the Christian world; and he who has erected one believes that he has atoned for every sin. But even the best are very miserable edifices of wattle plastered with mud, only to be distinguished from the surrounding hovels by a thin coating of whitewash, which is dashed over the outside to point with the finger of pride to the peculiar privilege of the two great powers in the land. Circular in form, with a door to each quarter of the compass, and a conical thatch, they have an apex surmounted by a brazen cross, which is usually adorned with ostrich eggs; and the same depraved and heathenish taste pervades the decorations of the interior. Sculpture is strictly forbidden; but the walls are bedaubed with paintings of the patron saint of the church, the Blessed Virgin, and a truly incongruous assemblage of cherubim and fallen angels, with the evil one himself enveloped in hell flames. Timbrels and crutches depend in picturesque confusion from the bare rafters of the roof, and no ceiling protects the head from the descent of the lizard and the spider." † Mr. Hormuzd Rassam has given a sketch of the ground plan of an Abyssinian church, showing the entrances and "courts." ‡ According

* "The British Mission to Abyssinia," vol. i. p. 227. Cf. p. 107 of this book.

† "The Highlands of Ethiopia," vol. iii. pp. 133, 134.

‡ "The British Mission to Abyssinia," vol. i. p. 217.

to him "the outer enclosure is for the congregation generally, male and female, the former occupying the western portion, and praying with their faces towards the east; the latter standing opposite to them and facing the west." *

The "crutches" mentioned by Major Harris astonished Mr. Vivian at a later date. He saw them in a monastery, and "imagined they had been contributed by lame people, to whom the saint of the place had restored the use of their legs. As a matter of fact, however, they were merely the crutches which every priest uses in the ritual of the dance." † The manner in which these singular adjuncts of the priestly office are employed has been described by Mr. Vivian. He attended a service at Trinity Church in Addis Abiba, and has given a very interesting account of the ceremonies. At a certain stage "all the priests grasp their long crutches and go through a kind of gymnastic exercise, which reminds me of the use of spears at a Somali dance. The crutches are five feet long, and must, I imagine, have originally taken their origin from spears, adapted for civilian use. The tops are either of ivory or brass, some of them elaborately carved. . . . One priest acts as conductor, and the others imitate his movements, all singing loudly through their noses. . . . The crutches are held in the middle and darted at the ground, now near, now far, with a forward movement made by slightly bending the right knee. It is as though spears were being poised and aimed playfully at objects on the floor. The crutches are then lifted, crook end up, a foot into the air, they are poised, they are swung, with ever-increasing vigour. All of a

* "The British Mission to Abyssinia," vol. i. p. 217. Stern did not observe this separation of the sexes.

† "Abyssinia," p. 277.

sudden the whole exercise ceases without any warning whatever." *

Major Harris wrote, with regard to certain other clerical functions—"The Father Confessor is bound to the strictest secrecy; and it is believed that on this point a dread oath is taken before ordination, when all the mysteries are expounded by the Abouna, and especially those which have reference to the preparation of bread for the holy supper. In a small house styled Bethlehem,† which rises immediately behind every church, the mysterious ceremony is performed. The deacon can alone bake the cake; and the most vigilant guard is invariably preserved against the approach or intrusion of females or other improper visitors during the hour of solemn preparation." ‡

The first appendix to Vol. III. of Major Harris's book contains a list of the sacred writings used by the Abyssinian Church. These include, *inter alia*—

The Old Testament.

The four gospels with readings, and all the other books of the New Testament.

Genset. A book used in funeral solemnities, ascribed to Athanasius, and stated to have been discovered by Helena at the digging out of the Holy Cross.

Retna Haimanot. The orthodox faith.

Henoch. The prophecies of Enoch.

Gadela Tekla Haimanot. Life of Tekla Haimanot. (And the lives of many other saints.)

Synodos. Canons of the Church, attributed to the Apostles.

Sena Febrak. History of the Creation; containing certain fabulous traditions concerning the Creation and the Antediluvian world,

---

* "Abyssinia," p. 288.

† Cf. p. 224.

‡ "The Highlands of Ethiopia," vol. iii. p. 132.

said to have been communicated to Moses on Mount Sinai, but not recorded in the Book of Genesis.

Tamera Mariam. Miracles of the Holy Virgin, wrought during her sojourn in Abyssinia, where she is said to have tarried three years and six months with the infant Jesus, before her return to Palestine.

Nagara Mariam. Words of the Holy Virgin.

Ardeet. Words said to have been spoken by Christ before his ascension.

Kedasie. Liturgy of the Abyssinian Church.

Gadela Adam. History of Adam.

Auda Negest. Book for prognostication.

Kufalik. Mysteries revealed to Moses on Mount Sinai, not written in the Pentateuch.

Serata beita Chrestian. Institutions of the Christian Church.

Degwa. Book of anthems, in which all the pieces of the liturgy that are chanted are set to music by St. Yareed, a native of Simien, who lived thirteen centuries ago and is believed to live still.

Lefafa Zedik. Prayers and spells against evil spirits and diseases, a book much esteemed, and buried along with the corpse.

Zalota Musa. Prayers of Moses against the influence of evil spirits.

Melka Michael. Prayers to St. Michael.

Mazafa Tomar. A letter which Christ is said to have written.

Germama. Prayers to frighten evil spirits.

Fekaric Yasoos. Christ's prophecy of the consummation of the world.

Haimanot Abao. Doctrines of the Abyssinian Church, comprising extracts from the Holy Scriptures, from synods, councils, and writings of the Fathers.

Gelota Monakosat. Prayers of the monks.

Felekisus. Book on monastic subjects.

Gadela Yob. Life of Job.

Raia Miriam. Dream of the Holy Virgin.

Gadela Samoetal. Lives of the martyrs.

The second appendix of the same volume contains a summary of the "Senkesar, or Synaxaria, being a *Collectio Vitarum Sanctarum* together with the Calendar of the Ethiopic Christian Church." The compiler of the summary wrote, "The following calendar, translated from the Latin of Ludolf, has been considerably enlarged by a comparison at Ankóber with a complete copy of the 'Senkesar.' The lives of the Saints, or the detail of miracles written against each day, are publicly read in the churches at the services beginning at the cock's first crowing."

The first month of the Abyssinian year is September. It is not surprising that in the long list of saints, patriarchs, prophets, martyrs, virgins, hermits, kings, "Fathers," presbyters, emperors, and male and female ascetics whose deeds are commemorated between New Year's Day (September 1, by Ethiopian reckoning) and the festival of St. Andrew and Moses, Bishop of Ferme, (being the thirtieth and last day of August according to the same computation), there should occur many names of worthies who are unknown to most European readers. For instance—

Abba Anbasa, who rode upon a lion.

Eustathius, who brought a dead child back to life. (This miracle is attributed to many of the saints.)

Ananias, who baptized St. Paul.

Pantaleon.

Jemrah, a martyr. On the spot where his blood was spilt there grew up a fine vine.

Naakweto-Laab, the last Emperor of Ethiopia, of the family of Zague. He did not die.

Cosmus, Metropolite. It is said that the image of Mary shed tears when it beheld his excruciating tortures.

The two hundred and ninety-two brothers and forty-nine sisters of Cosmus.

Abba Marina, who was discovered after death to have been a woman.

Barsufius.

Johannes Cama, whose fingers and nails seemed burning like candles during prayer.

Abba Libanus, who brought water out of a rock.
Daniel, the woman-hater, who vowed never to look at a woman.
The forty-nine old hermits.
Joseph, who was saved out of a fiery oven.
The 80,107 martyrs with Isidorus.
Atom and Arianus.
Eusebius, who, "being threatened to be burnt, was taken to heaven by Uriel the archangel, and remained there fourteen years." *

And many others whose sanctity is as little known in the Western Churches.

If there is anywhere a leisured and adventurous student of comparative hagiology, he would obtain some interesting results by tracing the origin and growth of the legends which glorify these worthies.

A survey of what has been said shows that Consul Plowden was fully justified when he wrote, "the Christianity professed and taught in Abyssinia is much materialized." Nevertheless, the credulity and superstition which prevail in the country are not in essence grosser than the unwarranted beliefs which were accepted in Europe when the *Aurea Legenda* passed as sound devotional reading, and it would be as unjustifiable to judge the intellectual capability of the Habashes by their faith in myths as it would have been to estimate the ability of our forefathers by a similar standard. Nor is it certain that any sudden reformation would improve the religious state of the people. An intemperate zeal for enlightening them might, by introducing an ideal far above their real sympathies and comprehension, discredit the old conceptions without effectually substi-

* This appendix of Major Harris's book was evidently prepared with care, and it is therefore surprising to find the saints, whose deeds are commemorated on October 4, described as "Papa and Mamma"! Presumably the explanation is that there was humour among the compositors of the year 1844, and that the compiler neglected his proof-reading.

tuting the new. Moreover, the fabulous feats of the saints may require a cautious and sparing criticism; for to differentiate the degrees of probability in miracles, among an unscientific but argumentative race, would, perhaps, be more likely to promote scepticism as to all thaumaturgy than to lead to the repudiation of certain miracles and the continued acceptance of others.

It is, of course, the fact that at present "religion" consists of little more than a mere series of outward observances. Veneration of the fabric of the church takes the place of devotion within its precincts. Dufton wrote:—"The prayers are read in Ethiopic" (Geez), "a language which the people know nothing about, so that little profit can be derived from the service. Indeed, most persons content themselves with kissing the floor or the walls of the edifice, and such is a criterion of a man's piety; he kisses the church, they say, and so esteem him a good Christian. Some will utter a prayer. The petition takes a form similar to the following, which an old woman was heard to offer up during my visit, though the last clause is probably in most cases omitted—

"'O Lord, give me plenty to eat and drink, good raiment, and a comfortable home, or else kill me outright!'"

Mr. Vivian visited the Church of the Blessed Virgin at Entotto, and noticed that all the Abyssinians who were with him "hastened to kiss the floor-beam and side-posts or lintels of the door in the outer wall." And at Trinity Church in Addis Abiba "whenever a peasant approached the church on a mule, he dismounted and kissed a large black stone which stood some twenty yards away. Other persons of a more pious turn of mind went right up to the entrance and kissed not only the lintels, but even the floor-beam of the doorway. This was also done by everybody who was coming to church, and my servant, in whom I had never detected any semblance of piety on the road, was particularly punctilious in this respect." *

But it is not the case that the Habashes avoid all religious duties that are irksome. Parkyns wrote: "Their fasts are more numerous perhaps than those of any other Christian people, more than two-thirds of the year being assigned to abstinence. Nor in their fasting do they get off as easily as Roman Catholics; for it is not sufficient that they should abstain from animal food only; an Abyssinian, during fast-time, neither eats nor drinks anything till late in the afternoon. . . . It is true the Mohammedans do nearly the same during their month of Ramadan; but they only change the day into night, feasting during the night-time on more luxurious food than many of them could allow themselves during the remainder of the year; while the Abyssinian, when he does eat, confines himself to dried peas, dressed in a sort of bad oil, or to an equally unpalatable dish made of a kind of spinach, called 'hamly' or 'goummen.' . . . Many of the Abyssinian fasts are of long duration. The time of day when the people may eat is determined by the length of a man's shadow, measured by his own feet, and varies in different fasts. Thus the fast of Advent is during the last ten days of the month Hedar (October) and the whole of Tahsas (November), and during each day till a man's shadow measures nine and a half feet. The fast of Lent lasts till sunset during fifty-five days. . . . Besides these are the Wednesdays and Fridays, making nearly 260 days of fasting. . . . Some of the priests

* "Abyssinia," pp. 280, 285.

are very rigorous in keeping all these fasts, and many even voluntarily add a number for their own observance. The people, too, in general are tolerably attentive to this duty; and I have frequently met with men undergoing extreme labour, yet persevering in what they have been brought up to consider as one of the most essential parts of their religion; for, strictly speaking, a man who has been known to neglect the rules of the Church is looked upon almost as an infidel, and should he die in such a state of disobedience, his body would be refused sepulture in the churches. Good Friday and the following day are passed by the priests and the rigidly devout in an absolute fast of forty-eight hours." *

Plowden remarked: "So much do they attach importance to this (fasting) and other outward forms that a man of Hamazayn † will slay his near relative, and returning home calmly, will be horror-stricken should his wife have ground flour on a saint's day, or prepared his meal before the hours of fasting have expired." ‡

It must be owned that the Habashes recompense themselves as far as possible for the severity of their abstinence by gluttony at the numerous feasts which the Church also ordains.

Major Cornwallis Harris drew attention to a singular custom in the following words—"Fasts, penances, and excommunication form the chief props of the clerical power; but the repentant sinner can always purchase a substitute to undergo the two former, and the ban of the Church is readily averted by a timely offering." § This seems to indicate an opinion among the Abyssinians that a sin is expiated not by the contrition of the evil-

* "Life in Abyssinia," pp. 277, 278. † A province of Abyssinia.

‡ "Abyssinia Described," by J. C. Hotten, p. 160.

§ "The Highlands of Ethiopia," vol. iii. p. 139.

doer, but by a formal amount of punishment undergone somewhere by some one—no matter by whom.

The Abyssinian Church teaches, and the great majority of the people accept, the doctrine of transubstantiation. Those who hold this belief "assert that the actual body and blood of our Saviour are partaken of by the faithful, but that an angel takes them away from an unbeliever, and restores the bread and wine, in his hands, in their natural state, such as they were previous to the benediction. The wine is merely an infusion of dried raisins." * Stern tells the story of a miracle which is supposed to have established the fact of transubstantiation, and adds, "The erudite reject the legend, and, in their sentiments, approximate to the Lutheran doctrine of consubstantiation." † The Church also teaches the doctrine of purgatory and that "prayers for the dead are necessary, and absolution indispensable. The souls of the departed do not immediately enter upon a state of happiness, the period being in exact accordance with the alms and prayers that are expended upon earth." ‡ Stern declared that "the number of masses requisite for the repose of the soul has not been defined by the Church, and thus the misery or bliss of the defunct is at the mercy of niggardly relatives and exacting priests." § Both writers attribute the introduction of this opinion to the Jews.

The badge of Ethiopian Christianity, worn by all classes, is the *mateb*, a blue neck-thread of silk or wool. This is the distinctive mark of the Abyssinians' creed; they hardly believe that one who does not wear it is in

* Parkyns, "Life in Abyssinia," p. 294.

† "Wanderings among the Falashas," p. 218.

‡ Harris, "The Highlands of Ethiopia," vol. iii. p. 141.

§ "Wanderings among the Falashas," p. 181.

reality a Christian, and, said Dufton, besides disdaining their Mohammedan and Jewish neighbours, "they often include in their contempt every form of Christianity even which does not conform to their own" *—in which prejudice they are not so singular as at a first glance they may appear to be. Commonly a silver ring or cross is attached to the cord.

The general character of the clergy appears from what has already been related. Parkyns saw them in remote places and in their least spiritual aspect. For instance, in describing the feast on St. Michael's day at Addy Harisho, he wrote: "An old priest came up to me and offered, on the part of himself and his brethren, to perform, if I pleased, the religious dance and song used by them on such occasions. I never shall forget their ludicrous efforts to appear graceful, at the same time staggering every step; while the expression of devotion they affected to assume was reduced to a languid smile and thickening eyelids, expressive of nothing but liquor. A hiccup or two occasionally interfered with the solemn words they were chanting; and the stately movements they had begun with changed gradually, and by degrees the dance became a reel, or rather reeling movement, the words only which accompanied it remaining solemn. At last an old priest, suddenly forgetting the original chant, changed its words to those of a jovial drinking ditty: 'Don't you stop the liquor, and I will dance for ever!' Instead of the marks of disapprobation from his fellow-priests, they only burst into a loud laugh, and, declaring the entertainment to be changed for the better, all with one consent followed his example and his tune." †

* "A Journey through Abyssinia," p. 91.

† "Life in Abyssinia," p. 163.

Nevertheless, Parkyns wrote of the priests with tolerance and good humour. And it is worthy of note that two other British writers who knew the country well have had a good word to say for the clergy. Plowden, after mentioning some of their faults, observed: "It is just to say that they have preserved the Christian faith, impure indeed, but still alive, in the midst of foreign invasion, domestic degradation, and the extinction of Government, and that it is under their protection that agriculture flourishes, and villages are built where deserts would else be seen." * Mr. Augustus Wylde, among several remarks in a similar sense, said: "I have found the clergy, if left alone, peaceable, simple-minded men, very hospitable and always willing to do me a good turn, and ready to help me and pass me on to their neighbouring friends, and I expect other travellers would find them the same as long as they treated them properly." †

An interesting instance of the survival of a Jewish institution among the Christians of Abyssinia is that of Cities of Refuge. Plowden wrote, "All the larger towns are entirely under their (the priests') control, and being cities of refuge, sacred even from the Ras, are filled with dissolute and dangerous characters." ‡ Parkyns mentioned Axum as a place of this kind, and added, "at Adoua the whole of that part called after the church, in fact the parish, is sanctuary ('guddam'), and no person having taken refuge there can be arrested, although he walk about the public streets in broad daylight, so long as he does not pass the parish boundaries." § Stern alluded to the subject, and had formed the impression that churches in general were sanctuaries. "In most cases the murderer may elude the violent rage of his pursuers

* "Abyssinia Described," p. 164.

† "Modern Abyssinia," p. 164.

‡ "Abyssinia Described," p. 159.

§ "Life in Abyssinia," p. 91.

by taking refuge in a church, where the priests will negotiate the price of his release." * He mentioned also certain rivers—the Taccazze and the Abai—forming the limits of provinces, which the "avenger of blood" may not pass.†

The Jewish Sabbath is still held in veneration by the Abyssinian Christians. Parkyns observed that a feast called "adam Souaur," or "left Saturday," was observed, "as that day was formerly the Jewish Sabbath, but given up by the Christians for the Sunday at the time of the resurrection of our Saviour. The priests collect a quantity of rushes, and bless them in the church; and then a priest and a deacon take them out into the parish, the deacon carrying the bundle, and the priest distributing the rushes to all passers-by, binding them on the head of the receiver. They also visit the houses with them, and generally receive a handsome present." ‡ It would be interesting to discover the origin of this custom. Major Harris, writing in 1843, said, "The Jewish Sabbath is strictly observed throughout the kingdom. The ox and ass are at rest. Agricultural pursuits are suspended. Household avocations must be laid aside, and the spirit of idleness reigns throughout the day." § The same writer has given a brief history of the observance.‖

It is well known that at one time Abyssinia was, so far as the Emperor's power extended at that period, converted to Roman Catholicism by the instrumentality of Portuguese missionaries. The connection with Portugal commenced in 1499, when Pedro Covilham penetrated by way of the Mediterranean and the Red Sea, to Shoa,

* "Wanderings among the Falashas," p. 131. † Ib.
‡ "Life in Abyssinia," p. 289.
§ "The Highlands of Ethiopia," vol. iii. p. 148. ‖ Ib.

where the Emperor then held his court. He was well received. An ancient Ethiopian law decreed that no foreigner who reached Abyssinia should be allowed to leave the country, but Covilham was permitted to send home an account of his experiences. At a later date a further communication from the traveller induced the Portuguese Government to despatch an embassy, which, after many privations and adventures, arrived in Shoa in October, 1520. It brought no presents, and was not cordially welcomed. However, the Emperor desired to send an answer to the King of Portugal, and therefore—in defiance of the ancient usage—allowed the ambassador to return to Lisbon, taking with him an Ethiopian envoy. But certain of the Portuguese were detained in Abyssinia, among whom was a priest named John Bermudez.

Some years later this man was appointed Abouna, or Patriarch, on the death of the incumbent of that office. He made his way to Rome, where Pope Paul III. confirmed his title, and he thence proceeded, in accordance with the Emperor's desire, to Portugal, and solicited the aid of European troops to assist the Abyssinians in repelling the Mohammedan invasion which then threatened the country.

The Portuguese Government despatched an expedition of some four hundred men under Don Christopher de Gama. These troops disembarked at Massowah in 1542. In an engagement with the Mohammedan forces on August 30 of that year the Abyssinian allies took to flight, and the Portuguese were surrounded. They cut their way through the enemy and effected their retreat, but Don Christopher, who had been wounded, was captured by the enemy and murdered by the Mohammedan commander. His skull was sent to Constantinople as a trophy. The surviving Portuguese besought of the Emperor an opportunity to avenge their leader, and they shortly afterwards took part in an action in which the Mohammedan forces were routed with great slaughter.

These circumstances added to the influence of Bermudez, but he insisted rigorously on the complete acceptance of Roman doctrine and ritual by the Abyssinians, and thus incurred the resentment of the Emperor. As a consequence he lost the advantages which he and his compatriots had gained, and finally abandoned his task and quitted the country, embarking at Massowah, where he had landed.

A subsequent mission under the leadership of Oviedo, Bishop of Hieropolis, which reached Massowah in 1558, failed equally, after many adventures—including an Abyssinian Council at which the Emperor and the bishop publicly and heatedly disputed about points of doctrine—and in 1560 Oviedo received orders from Rome to withdraw from Ethiopia and proceed to Japan.

Peter Paez was the next missionary who attempted to bring Abyssinia into communion with the Catholic Church. After suffering shipwreck in the Red Sea and enduring a seven years' captivity in the hands of the Turks he landed at Massowah in 1600. Paez was a man of prudence and ability. He prepared himself for his task by learning the Geez language and opening a school at Maiguagua, in which he was able to study the character of the Abyssinian people. He acquired a reputation as a teacher and disputant, and in 1604 was summoned to the court, where the reigning Emperor, Za Dengel, received him as a person of distinction. He soon acquired influence over this monarch, who sanctioned the use of the Roman ritual and shortly became a convert to Catholicism. A rebellion—one of the ordinary incidents of Abyssinian politics—was organized against him, and he was slain. A period of civil war ensued, during which his successor shared the same fate.

The crown then passed to the Emperor Segued (called also Socinios and Susneus).* Paez aroused his interest by his skill in architecture and his other talents, was received into favour, and for the second time induced the ruler of the country to recognize the authority of the Catholic Church. Letters announcing the event, signed by the Emperor himself, were sent to the Pope and the King of Portugal. Paez died in 1623, shortly after receiving Segued's formal abjuration of heresy and administering to him the sacrament of penance.

Upon receipt of the Emperor's letter, the Pope, in 1624, despatched to Ethiopia a mission headed by Alphonso Mendez, a Jesuit and doctor of divinity. This envoy and his followers reached Abyssinia in 1625. On February 11, 1626, before the court and the magnates of the country, the Emperor avowed his submission to the head of the Catholic Church, and took the oath of obedience to the Holy See. A proclamation was then made that "all persons intended for priests should embrace the Catholic religion under pain of death, and that all should follow the forms of the Church of Rome in the celebration of Easter and Lent under the same penalty."

The change of creed was not popular in the country. It was opposed by force, and became the occasion of a renewal of civil war. The Emperor found that he had alienated his subjects, and, in spite of the protests and

* This was the "Sultan Segued" who caused the still existing bridge over the Blue Nile to be built. See p. 147.

resistance of Mendez, he issued, in 1632, an edict by which he restored the ancient religion to its former authority, and abdicated in favour of his son. Segued died a few months afterwards, still professing the Catholic faith. His successor banished Mendez and his retinue, whereupon they sought protection with the Prince of Bar, who had been at enmity with the Abyssinian king; but this chieftain delivered them to the Turks. Father Jerome Lobo, who wrote a history of the mission, and some of his companions, reached Europe after many adventures; others were ransomed from the Pasha of Souakim and proceeded to India, and the few Jesuits who had remained in Abyssinia were put to death.

Of nine Capuchins who endeavoured to enter the country soon after these events, seven were killed, the three who arrived last being murdered by the Pasha of Souakim, at the Emperor of Abyssinia's request, as soon as they reached the coast.

No subsequent attempt of similar importance was made to convert the Abyssinian nation to Catholicism, and the story of the struggle between the Roman and the Alexandrian forms of belief may be said to have ended with the banishment of Mendez.

There have been theological contentions in the Abyssinian Church as in others, and it has not escaped schism. The fiercest dispute arose about the number of the births of Christ, and the sword has been used—as elsewhere—to lend force to arguments. Major Harris wrote: "At the expense of a bloody civil war, Gondar, with Godjam, Damot, and all the south-western provinces of Amhára, has long maintained the three births of Christ—Christ proceeding from the Father from all eternity, styled 'the eternal birth;' His incarnation, as being born of the holy Virgin, termed His 'second or temporal birth;' and His reception of the Holy Ghost, denominated His 'third birth.' The Tigre ecclesiastics, on the other hand, whose side is invariably espoused by the Primate of Ethiopia, deny the third birth, upon the ground that the reception of the Holy Ghost cannot be so styled." * Bishop Samuel Gobat, whose book was published in 1834, gave a fuller account of the controversy, which Parkyns quoted.† In Stern's time the Abouna (primate) was supported by King Theodore in his opposition to the doctrine of the three births, and "the royal herald made proclamation that in future all who adhered to the obnoxious dogma of the threefold birth would be taught obedience by the *giraffe* (scourging). The Shoa clergy denounced this decision as arbitrary and tyrannical, as indeed it was; but an application of the promised whip wrought a wonderful change among that insubordinate body." ‡ The same author on one occasion saw about a dozen priests in chains, who were brought before the Abouna to be admonished because "they had pertinaciously clung to the abolished dogma of the three births of Christ." § The penalty which they incurred "consisted of several months' successive fasts, divers fines, and the promise of the *giraffe*."

For a detailed account of Abyssinian Christianity and ecclesiastical history, reference should be made to the books already mentioned in this volume—Bruce, Gobat, Harris, Lobo, Legrand, Stern, Dufton, Plowden, and Parkyns especially—and to the works of F. Alvarez ||

* "The Highlands of Ethiopia," vol. iii. p. 186.

† "Life in Abyssinia," p. 295.

‡ "Wanderings among the Falashas," p. 306. § P. 136.

|| "Narrative of the Portuguese Embassy to Abyssinia," 1520–27, translated and edited by Lord Stanley of Alderley, Hakluyt Society, 1881.

and Job Ludolphus.* A digest of the writings of the early travellers and missionaries in Abyssinia was published by F. Balthazar Tellez, and translated into English by John Stevens (1710).

Mention has already been made of the Falashas (Jews) of Abyssinia. These people, said Stern, "claiming a lineal descent from Abraham, Isaac, and Jacob, pride themselves on the fame of their progenitors and the purity of the blood that circulates in their own veins. Intermarriages with those of another tribe or creed are strictly interdicted,† nay, even the visit to an unbeliever's house is a sin, and subjects the transgressor to the penance of a thorough lustration and a complete change of dress before he can return to his own home. Their stern uncompromising sectarian spirit has been highly beneficial in excluding from their community that licentious profligacy in which all the other inhabitants of Ethiopia riot; and it is generally admitted that Falasha men and women seldom, if ever, stray from the path of virtue, or transgress the solemn law of the decalogue." ‡

Various accounts, of which two have already been mentioned, are given of the origin of the Jewish colony in Abyssinia. It is commonly believed among the people that a large retinue returned with Maqueda, Queen of Sheba, from her visit to King Solomon. Harris remarked, "The tradition of Queen Maqueda has been

* "History of Ethiopia," made English by J. P. Gent, 1684.

† The reader need hardly be reminded that the same interdiction is in force elsewhere. It is the refusal of the *connubium* that renders the British population in the East of London and elsewhere unable to assimilate the large colonies of Jews who have lately immigrated into this country, and in great part causes the difficulties of the "Alien Question." The subject has been treated in Major Evans Gordon's very interesting book, "The Alien Immigrant" (1903).

‡ "Wanderings among the Falashas," p. 187.

ascribed to the invention of those fugitive Jews, who, after the destruction of Jerusalem by the Emperor Titus, emigrated into the northern States by way of the Red Sea, who disseminated it with the design of obtaining the desired permission to settle in the country, and whose descendants are the Falashas still extant among the mountains of Simien and Lasta."* According to Stern, "the most probable conjecture is that at a very early period—perhaps when Solomon's fleet navigated the Red Sea—some adventurous Jews, impelled by love of gain, settled among the pleasant hills of Arabia Felix; whilst others of a more daring and enterprising spirit were induced to try their fortune in the more remote mountain scenes of Ethiopia. . . . Subsequent troubles in Palestine and the final overthrow of the Jewish monarchy by Nebuchadnezzar increased the number of the emigrants, and in the lapse of a few centuries the Jews formed a powerful state in Arabia, and a formidable and turbulent people in the Alpine regions between Tigre and Amhara in Ethiopia." † Mr. Vivian alluded to "a colony of aboriginal Jews up in the mountains of Tigre. They live in pastoral fashion, like the old Hebrew patriarchs, upon the produce of the flocks and herds. They have been there for centuries, perhaps even for thousands of years, and the Abyssinians confess that they have always failed to dislodge them from their inaccessible fastnesses." ‡ Falashas are also found in the provinces bordering on Lake Tsana, including Godjam, and it was in this district that Stern visited them. It seems probable that the Jews who have retained their religion are the descendants

* "The Highlands of Ethiopia," vol. iii. p. 145.

† "Wanderings among the Falashas," pp. 185, 186.

‡ "Abyssinia," p. 329.

of those who refused to conform to the doctrines of the Coptic Church when they were generally adopted in the country. Bruce mentions that "about one hundred and eighty years after the establishment of Christianity, a religious war is said to have taken place between the converted and unconverted Abyssinians (the Christians and the Jews)," * and if one may judge by the steady and successful opposition which Hebrews in other countries have offered to the most strenuous methods of conversion, it appears likely that a portion of those in Abyssinia adhered to their ancient creed.

There is a tradition that in the tenth century the Jews obtained ascendency in Ethiopia, and preserved their supremacy for about three centuries. "In the year 960 the Jews, supported by their king and by his daughter Judith, a woman of great beauty, resolved to attempt the subversion of the Christian religion and the destruction of the race of Solomon (Queen Maqueda's descendants having been converted to the Alexandrian doctrines). They surprised the mountain of Damo, the residence of the Christian princes, the whole of whom, about four hundred, were massacred, excepting one infant, who escaped into the powerful and loyal province of Shoa. A solitary representative of the blood of Solomon and the Queen of Sheba was thus preserved. Judith took possession of the throne, and not only enjoyed it herself for forty years, but transmitted it to five of her posterity. On the death of the last of these the crown descended to one of his relations, a Christian, and is said to have remained in his family (who, though Christians, were not of the line of Solomon) for five generations. Finally, Tekla Haimanot, the famous saint, persuaded the reigning king to restore the crown to the

* Major Head's "Biography," p. 162.

house of Solomon, whose descendants had survived in Shoa." *

Differing accounts of the former political status of the Hebrews in the land have been given by various writers; Stern said that in their "mountain fastnesses of Simien and Bellesa they maintained, under their own kings and queens, called Gideon and Judith, a chequered and independent existence till the beginning of the seventeenth century." † Harris, following Bruce more closely, wrote, "Christianity became the national religion of Abyssinia in the beginning of the fourth century. The Falashas, descendants of the Jews who were believed to have accompanied Menelek from Jerusalem, had meanwhile waxed extremely powerful, and refusing to abandon the faith of their forefathers, they now declared independence. Electing a sovereign of their own creed, they took possession of the almost impregnable mountain fastnesses of Simien, where their numbers were augmented by continual accessions from the Jews who were expelled from Palestine and from Arabia. Under the constant titles of Gideon and Judith, a succession of kings and queens held a limited sway until, in the middle of the *tenth* century," the Princess Esther brought about the revolution which Bruce ascribed to Judith—one writer probably using the name and the other the title of the usurper in question. Whatever the circumstances may really have been, there seems reason to believe that the Jews were at one time sufficiently influential to impose a dynasty of their choice upon the nation.

Stern has given a full and very interesting account of the Falashas whom he saw. According to this they

* Major Head's "Biography of Bruce," pp. 162, 163.

† "Wanderings among the Falashas," p. 186.

dwell as a rule, though not invariably, apart from the remainder of the population, and "their settlements are strikingly distinguished from the Christian villages by the red earthen pot on the apex of their *mesquid,* or place of worship, which towers from the centre of the thatched huts by which it is environed. . . . Husbandry and a few simple trades—such as smiths, potters, and weavers—constitute the sole occupations in which they engage; commerce they unanimously repudiate as incompatible with their Mosaic Creed, and it is quite a disappointment not to find a single merchant among a quarter of a million of people, the lineal descendants of those who are supposed to have acquired a taste for traffic and riches on the very eve of their emancipation from Egyptian servitude."

Though they had very few copies of the Scriptures in Stern's time they were well acquainted with the Old Testament, especially the Pentateuch and the Psalms. "Their sacrifices," he wrote, "are most capriciously offered, and with the exception of the Paschal Lamb, neither the offering on the Sabbath nor on the day of atonement is in accordance with the original command. . . . Every Falasha settlement has a hut at its outskirts, and there the unclean and impure must take refuge during the prescribed number of days. This ritual scrupulosity involves many social hardships and inflicts on numbers many a keen pang. Particularly in the hour of dissolution, when the sweet expressions of friendship and love are so soothing to the agonized soul and anguished frame, the dying Falasha has no affectionate hand clasped in his, and no words of comfort from beloved objects whispered in his ears. The inflexible law forbids the last offices to the weeping relative, and the helpless sufferer is in death's agonizing convulsions dragged from the weary couch into the open air, where the polluted and unclean remove him from the bare ground to the tainted and lonely hut."

On the Sabbath "the service, which consists in chanting psalms and hymns relieved by allegorical stories and a few verses or a chapter of the book of Leviticus, lasts a considerable time, and in some places the plaintive notes of the worshippers may even be heard across the quiet valley and around the lonely hill throughout the night. . . . Broad phylacteries and the garments of fringes are utterly unknown among them, nor do they wash the cup or practise any of the decrees of the rabbins. . . . About the advent of the Messiah they have no intelligent or definite idea. 'We believe that Jerusalem will again be rebuilt,' is the answer on the lip of every Falasha, when questioned as to the future destiny of his nation. This event they regard as the consummation of their brightest hopes—the realization of their fondest mundane visions. . . ."

Asceticism, borrowed from the Christian hermits, has become a practice among the Falasha priests, who "after their initiation frequently pass months and years in swampy marshes, stern wilds and poisonous jungles, where roots or dried peas, which latter they carry with them, are their only means of subsistence. Numbers succumb to the noxious influence of the atmosphere, others perish of famine, whilst not a few become the prey of the lion, hyena, and other voracious beasts which inhabit those unsightly tracts." There are also Falasha monks, and "the dwellings and convents of these ascetics are carefully isolated from the abodes of the impure and unholy people."

There were, in Stern's day, three Falasha high-priests in Abyssinia, one in the province of Quara, the other in Armatgioho, and the third, Aboo Maharee, in Dembea. The missionary had some indecisive discussion with the last-named, a man about sixty years of age, "of a noble and commanding figure, swathed in a white *shama*, and holding a long bamboo staff, which in the distance looked like the crosier of a bishop." He had a "high and expressive forehead, melancholy restless eyes, and a countenance once no doubt mild and pleasing, but to which self-imposed penances and a repulsive practice had imparted an expression most strange and unearthly."

"The common people have all an erect, upright carriage;" the women are intelligent and *sympathiques*, and the younger of them attractive. To judge by Stern's description,* the Falashas are generally a worthy, simple and devout race, bigoted but "exemplary in their morals, and cleanly in their habits"—in which latter respect they form a striking contrast to their co-religionists in the slums of London. Plowden said of the Abyssinian Jews, "They are still found in some numbers, and, though despised, are not persecuted; this may be owing to their poverty. They know nothing of the Hebrew tongue, but some read the Mosaic books in Geez, and are as scrupulous in their ceremonials as their brethren elsewhere. They are the best masons in the country." †

"The *mesquids*, like the Christian churches, consist of three divisions, with an entrance towards the east. The admission into these different courts is rigorously regulated by Levitical law, and the severest penalty would be inflicted on any one who should incautiously pollute the sacred edifice. In the rear of every place of worship is a small enclosure with a huge stone in the

* See "Wanderings among the Falashas," chap. xiv. *et seq.*

† Hotten, p. 166.

centre; and on this crude altar the victim is slaughtered, and all other sacrificial rites performed." *

For a notice of a curious superstition of which a class of Jewish people are often the object among Christian Abyssinians, see p. 288.

In the "Jewish Year-Book" † for 1904 the number of Jews in Abyssinia is computed at 50,000; the "American Jewish Year Book" estimates it to be 120,000. Both these figures must be conjectural, and the former differs very widely from that given by Stern, who had had opportunities of obtaining information in the country. Considering the fecundity of the Jews in most parts of the world it is unlikely that their numbers have diminished in Ethiopia during the last forty years.

In Plowden's time there was a large and prosperous Mohammedan population in Abyssinia. He wrote: "In all large towns they have a separate quarter, with mosques and public prayers. From the advantage that their commerce in slaves gives them over their Christian competitors, the Mussulman traders are the most wealthy, and are, therefore, generally appointed to the high post of Negadeh Ras, or collector of all customs, literally 'head of merchants.' To enforce their authority these keep large bodies of armed men, and confidently predict the final triumph of the faith of the Prophet in Abyssinia. The Abyssinian Mussulmans, as distinguished from the Galla, are all traders; they will not eat meat killed by Christians, and are frequently their superiors in morality and intelligence. They live on terms of equality, good humour and friendship with the Christians, openly defend their creed, and receive any proselyte that offers, and do not appear to think that the restrictions

* "Wanderings among the Falashas," p. 188.

† Greenberg and Co., 80, Chancery Lane, London.

in the Koran respecting strong drinks apply to them at all." Plowden added that they, like the Christians, did not generally seclude their women.

King Theodore, early in his reign, issued an edict which "required peremptorily the expulsion from the country, or the instant return to the bosom of the Church of his apostate countrymen," *i.e.* Mohammedans.* Circumstances prevented him from enforcing this decree. It was revived by King Johannes, and no doubt, had in other parts of Abyssinia the effect which Stecker observed in Korata.† But Mohammedans are still to be found—*e.g.* in the district of Axum ‡ and in Shoa.§

Allusion has already been made to the Waitos.‖ Plowden wrote of them: "They call themselves Mohammedans, but are not recognized by the other followers of that creed. They principally reside near the Lake Tsana, and are a very handsome race. They are regarded with as much aversion as the Jews."¶ It must be owned that the appreciation of their personal appearance is a matter of taste.

Another singular sect is that known as Koomants, or Kamants. These people, "found only in the neighbourhood of Gondar, are acknowledged by neither Christian, Mussulman, or Jew, and have a bastard creed, a

* Hotten, pp. 167, 168.

† "Dufton, "A Journey through Abyssinia," p. 143.

‡ The ancient city of Axum is remarkable for the ruins and other traces of an early and extinct civilization which it contains. These have been the occasion of much speculation among archæologists. An account of the obelisks and other remains was given by Bruce; Mr. Wylde has made some interesting comments on the subject in his chapter on Axum; many writers on Abyssinia have referred to it at some length; and the late Mr. Theodore Bent dealt with the antiquities of the place in "The Sacred City of the Ethiopians" (Longmans, Green and Co., 1893).

§ Augustus Wylde, "Modern Abyssinia," pp. 18, 407.

‖ See p. 143.

¶ Hotten, p. 167.

compound of all three. They are skilful carpenters, and supply all Gondar with wood. They are despised, but being very courageous, and having lately shown an inclining towards Christianity, it is not improbable that their distinctions will soon disappear; many even now have ceased a practice which was the chief separating cause. They hung heavy weights in the lobe of the ear of the girls, who are thereby excluded from any chance of marriage with Christians." * Hormuzd Rassam on one occasion halted at a village of Kamants called Saraba, north of Lake Tsana, in the direction of Gondar. He saw there "a place of worship, which they hold in great veneration. The entire space between the spring and this temple, or, more strictly, grove—they have no building for their religious services—is regarded as sacred, and no Christian or Mohammedan is allowed to enter within the hallowed precincts. When engaged in prayer, they take up a position under a tree, on the banks of the stream, or in the wood about half a mile from the spring. It is owing to this practice that the Christian Abyssinians style them 'Worshippers of wood,' or simply 'Wood.' Their religion is as great a mystery in Abyssinia as that of the Ansairies is in Syria, and although the late king (Theodore) obliged them to wear the 'mateb' cord, yet they still continue to practise certain rites and ceremonies unknown either to Christians or Mussulmans. They only eat meat which has been slaughtered by themselves, and eschew fish and coffee; yet they adopt Christian names, and have a kind of baptism, which their Christian countrymen designate a 'mockery.'" †

* Plowden, quoted by Hotten, p. 166. Plowden's description applies to the fifties of the nineteenth century.

† "The British Mission to Abyssinia," vol. i. p. 209.

Stern's first experience of these people was agreeable. In a deep valley called Walee Dubba, on the way to Tschelga, he met some young women who had jars of *dallah,* a beer made from sprouting barley, to sell. "It was quite an unexpected surprise," he wrote, "to see the solitude of an African wilderness enlivened with the gay song and sprightly converse of a number of young lasses, who, clad in rustling leather petticoats, moved about amongst the various groups resting under the leafy foliage, with a grace and innate modesty which elicited admiration, whilst at the same time it forbad all unbecoming liberty." At Tschelga he questioned some of the Kamants "about their knowledge of God and their hopes of eternity; but they had so little to communicate, beyond a belief in a Supreme Being and the existence of a future state, that the most simple query caused them the utmost wonder and surprise. . . . Their language is Amharic, but amongst themselves they speak in the Falasha tongue; and the striking Jewish features of many a man and woman amongst them inclined us to credit the report which assigns to them a Jewish origin." The same writer has mentioned the legendary derivation of their name,* and he described them as industrious, energetic, and active. He saw many of their women in Gondar, and remarked that "it was a strange sight to see these young females—clad in simple leather petticoats and equally simple earrings of wood, which, according to the orthodox fashion, must be weighty enough

* "According to Abyssinian tradition, the King of Tigre, soon after his conversion to Christianity, crossed the Taccazze, and invaded Simien and Amhara. Here he met a people who were neither Pagans nor Christians, a marvel which aroused the monarch's curiosity, and he inquired what they believed; to which, in a laconic style, they replied in their own dialect, *Kam Ant, i.e.* 'as thou,' from whence they obtained their present appellation."

to distend in a few years the flap of the ear down to the shoulders—walking about in the market, or groaning under a heavy burden of wood, utterly unconcerned about everything except the graceful ornament that dangled round their necks." *

At the present time European creeds are represented in Abyssinia by converts, whose change of faith is due to the activity of missionaries. Mr. Wylde wrote upon this subject—"I believe the majority of the Abyssinians care a great deal for their religion, and it is only the more worthless ones that are found round the different mission stations; people who are willing to change their faith the same as they would their clothes, and when they have worn out all that are to be got, revert to their original one again, without perhaps being any the better or any the worse for the experience, but only to be marked by others as being utterly worthless and unreliable characters. I will never have a male servant in my employ that has been near a mission, if I can help it. Female servants are different; they usually are taught to sew, wash, and cook, and are generally cleanly in their habits, but when they get to a certain age the majority of them run away from these establishments, as they cannot stand the discipline and restraint; and I don't blame them, as a more unlovely and monotonous life does not exist." †

There is a code of law in Abyssinia, which is called *Fatha Negest*, or *Feth Negust.* According to Major Harris this volume of "The Judgments of the Kings" is said to have fallen from heaven in the time of Constantine the Great, and he includes it among the books of which

* "Wanderings among the Falashas," p. 42 *seq.*

† "Modern Abyssinia," p. 144.

MSS. are still extant in the country.* Plowden stated that it was supposed "to have been complied by the Council of Three Hundred, in the earlier ages of the Church, and regarded originally as of almost equal authority with the sacred writings."† He further remarked that this code "is a bad translation from that of Justinian; three parts of it being occupied with Church affairs and regulations, and a small portion only with the civil and criminal law, this latter being also much mixed up with the institutions of the Pentateuch. Bad as it is, there are probably not twenty persons in the country that are conversant with it; and some singular judgments are given on its authority, much after the fashion of the *sortes Virgilianæ*. Though it is consulted with much ceremony, and considered a sacred volume, law is, in fact, simplified to the will of the chief. Still, the fact of there being a written law has assisted in retarding the degradation of the people."‡

The following extract from Parkyns's work shows the spirit of Ethiopian jurisprudence in dealing with serious crime: "As for the laws of the country, they are for the most part formed on the basis of the old Mosaic dispensation. 'An eye for an eye, and a tooth for a tooth' is followed nearly to the letter—so much so that, if a man kill another, the murderer must be put to death by the nearest relatives of the deceased with precisely the same kind of weapon as that with which he killed his victim."§ The murderer may redeem his life by paying a ransom, which, in Stern's time, varied from fifty to two hundred and fifty Maria Theresa dollars.

* "Highlands of Ethiopia," vol. iii. p. 393.

† Hotten, p. 120. ‡ Ib. p. 186.

§ "Life in Abyssinia," p. 366.

If he did not possess the requisite amount he was chained to a relative of the deceased and obliged to beg till he had collected the stipulated sum. The severity of the punishments in vogue varies with the temperament of different rulers, the Emperor being the final authority in such matters, and a right of appeal to him is admitted in favour of all accused persons.* Mutilation is a common penalty.† "This," said Mr. Wylde, "has not the terrors that it would have in England, as some of the thieves in Abyssinia have been operated on a second and a third time, and I saw one man with his left foot being the only extremity left, and he was being fed by the priests at the church at Adowa." ‡

All Abyssinian litigants are required to find security. "These securities must be persons whom the judge is certain of being able to seize if necessary. If the charge be serious they must be persons of property, and are liable both for the appearance of their principals, and for the sentence whatever that may be. This custom obtains both in civil and criminal law. In default of such security, each party is chained by the wrist; an additional expense, as the chains must be hired, and the jailor—that is, the person to whom he is chained, paid a sum fixed by law. This bail is the prop of Abyssinian society; no commercial or market transfer takes place without it. The Abyssinian judge or creditor cares nothing for the principal in a cause; the bail is seized, and in self-defence produces his man; and it is an honourable trait that the principal rarely absconds. A friend will thus become security in a case of murder, though rendering himself thereby liable

* "Wanderings among the Falashas," chap. ix; Hotten, p. 121.

† See, *e.g.* "Life in Abyssinia," p. 368.

‡ "Modern Abyssinia," p. 310.

to suffer death, or to pay the price of blood, to them a fortune.

"After both parties have given security the plaintiff only is allowed to produce his witnesses; the defendant can, when they are called by name, admit or reject their evidence, in almost every case great indulgence being shown in this respect. . . . In small affairs, such as a sudden dispute on the high road, the meeting of an absconded debtor, or any civil matter, the first decent person to be found is obliged to act as a temporary judge, if adjured by the 'death of the chief' paramount. He must then place the accused in bonds, which is done by tying his cloth (*shama*) to that of his accuser, and escort or send them to the nearest magistrate, who, should the accused demand it, must in like manner forward him to his immediate master or chief, where the case is first heard, the plaintiff having right of appeal; the law in this being, however, highly favourable to the defendant, the plaintiff not being always disposed for a long journey." *

Every man is his own advocate. Those who wish to read an account of the proceedings "in court" should refer to "Modern Abyssinia," p. 308, or "Wanderings among the Falashas," pp. 48, 170. The hearing usually gives occasion for a good deal of eloquence and a great deal of lying.

There is no sort of general education in Abyssinia. Men of attainments are found in the upper classes, and some of the higher officials are capable linguists. But as regards the bulk of the nation learning has advanced little since the days of Harris. "The stores of literature being bound up in a dead letter (the Geez language), few excepting the priests and *defterers* can decipher them,

* Hotten, p. 183 *seq.*

and many of these learned men are often more indebted to the memory of their early youth than to the well-thumbed page in their hand. The ignorance of the nation is, indeed, truly deplorable; for those children only receive the rudiments of an education who are designed for the service of the Church; and, the course of study adopted being little calculated to expand the mind of the neophyte, a peculiar deficiency is presented in intellectual features." * Plowden remarked of the priests that they "teach but one book to the children of the laity, 'The Psalms of David;' and without forbidding other learning, discourage it, confining it as much as possible to the clergy and the scribes." † Allusion has already been made to the classes held by the priests in the precincts of the churches.‡ Such instruction as is obtainable is given, of course, to all pupils under clerical supervision. There are no secular schools, and at the present time, as Mr. Vivian remarked, "no one in Abyssinia ever does get any education beyond learning to read and write, and quote passages from the Gospels." § Those whose acquirements exceed this standard are not indebted for them to native teachers.

The habits of the young girls, among the generality of the people, do not foster any strict notions of morality. Parkyns wrote: "The boys are turned out wild to look after the sheep and cattle, and the girls, from early childhood, are sent to fetch water from the well or brook, first in a gourd, and afterwards in a jar proportioned to their strength. If the well be far from the village, the girls usually form parties to go thither and amuse themselves on the road by singing sentimental or love songs; while, during their halt at the well for an hour or so, they

* "The Highlands of Ethiopia," vol. iii. p. 184.

† Hotten, p. 158. ‡ See, *e.g.* p. 121. § "Abyssinia," p. 204.

engage in romps of all kinds, in which parties of the other sex frequently join. This early licence lays the foundation for worse, when at a later period they are sent to the woods to collect fuel." *

Courtship can hardly be said to exist. "In Abyssinia young people begin to think of marriage at a very early age. I have seen brides of eight or nine years old; and boys at a proportionally youthful age are considered marriageable. When a lad wishes to marry he only inquires for a girl who may possess twice the number of oxen that he can muster, or their value. His proposals are made to the girl's father, and, unless there is some strong motive for rejecting him, he is accepted, and everything is arranged without consulting the lady's taste or asking her consent. They are usually betrothed three or four months before marriage, during which time the bridegroom frequently visits his father-in-law elect, and occasionally propitiates him with presents of honey, butter, a sheep, or a goat; but he is never allowed to see his intended wife even for a moment, unless, by urgent entreaty or a handsome bribe, he induces some female friend of hers to arrange the matter, by procuring him a glance at his cruel fair one. For this purpose he conceals himself behind a door or other convenient hiding-place, while the lady, on some pretext or other, is led past it. Should she, however, suspect a trick, and discover him, she would make a great uproar, cover her face, and, screaming, run away and hide herself, as though her sense of propriety were greatly offended by the intrusion; although previous to his making the offer she would have thought it no harm to romp with him, or any other male acquaintance, in the most free and easy manner. Even after she has been betrothed,

* "Life in Abyssinia," p. 254.

she is at home to every one, except to him who most sighs for the light of her countenance.

"In Tigre, and especially in Shire, there is a superstition that if a girl leave her father's house during the interval between her betrothal and marriage she will be bitten by a snake.

"When the wedding day approaches the girl is well washed, her hair combed and tressed, and she is rendered in every way as agreeable as possible. She has then to undergo a course of diet and medicine." *

Of marriages there are three kinds. The foregoing extract refers to the usual merely civil contract; for such unions there are "bridesmen" (*arkees*), who have peculiar privileges in the community, including that of theft, remain "on duty" after the nuptials, and occupy the most singular relation to the newly wedded damsel.† There are also bridesmaids, who keep their faces covered during the wedding, which is a very simple ceremony. Mr. Wylde and an Italian friend obtained an illicit view of an Abyssinian bride and bridesmaids, and were rewarded by the sight of some very charming girls. Both he and Parkyns have written entertaining descriptions of an Abyssinian marriage of this kind.‡

Rassam has given the following particulars with regard to such contracts. "The parties simply swear, in the presence of two witnesses, that they will live together as husband and wife. This bond may be dissolved at any time by mutual consent; in that case, the wife is entitled to retain whatever property she possessed before wedlock, as also any presents which she may have received from her husband during coverture, unless a

* "Life in Abyssinia," pp. 255, 256.

† Parkyns, pp. 266, 267.

‡ "Modern Abyssinia," p. 181 *seq.* "Life in Abyssinia," p. 256 *seq.*

stipulation to the contrary was agreed to by both sides, on their union. Most Abyssinian marriages are of this sort, and the generality of the respectable classes so wedded live together as husband and wife until separated by death." *

Harris wrote: "In Shoa a girl is reckoned according to the value of her property; and the heiress to a house, a field, and a bedstead is certain to add a husband to her list before many summers have shone over her head. Marriage is generally concluded by the parties declaring, before witnesses, 'upon the life of the King,' that they intend to live happily together, and the property of each being produced is carefully appraised. A mule or an ass, a dollar, a shield and a sheaf of spears on the one side,† are noted against the lady's stock of wheat, cotton, and household gear; and the bargain being struck, the effects become joint for the time, until some domestic difference results in either taking up their own, and departing to seek a new mate.

"Favourite slaves and concubines are respected as much as wedded wives. No distinction is made betwixt legitimate and illegitimate children; and, to the extent of his means, every subject follows the example set by the monarch (Sahela Selassie), who, it has been seen, entertains upon his establishment, in addition to his lawful spouse, no fewer than five hundred concubines.

"Few married couples even live long together without violating their vow; and the dereliction being held of small account, a beating is the only punishment inflicted upon the weaker party. The jewel chastity is here in no repute; and the utmost extent of reparation

* "The British Mission to Abyssinia," vol. ii. p. 219.

† In matters of detail this account applies, of course, to Abyssinia in the first half of the nineteenth century.

to be recovered in a court of justice for the most aggravated case of seduction is but fivepence sterling.

"Morality is thus at the very lowest ebb; for there is neither custom nor inducement to be chaste, and beads, more precious than fine gold, bear down every barrier of restraint. Honesty and modesty both yield to the force of temptation, and pride is seldom offended at living in a state of indolent dependence upon others. The soft savage requires but little inducement to follow the bent of her passions according to the dictates of unenlightened nature; and neither scruples of conscience nor the rules of the loose society form any obstacle whatever to their entire gratification." *

Plowden confirms the observation that "no distinction is made between legitimate and illegitimate children," and adds, "all share alike on the decease of their father." †

The reader will have noticed that Rassam's account of the duration and even tenour of these alliances does not agree with the comments made by Harris. Stern observed that the wedded pair "may, perhaps, become attached to each other and live in peace and conjugal bliss; or, as it frequently happens, they may become disgusted with each other after the lapse of a brief period and separate." ‡

Parkyns wrote :—"The civil marriage may be dissolved on the shortest notice and for the most trifling reasons. Parties have only to express their wish to separate, and a division of property and children takes place. The boys, if there be any, usually go to the father, and the girls to the mother. This having been

* "The Highlands of Ethiopia," vol. iii. pp. 166-168.

† Hotten, p. 137.

‡ "Wanderings among the Falashas," p. 119.

done, the parties are at liberty to contract a second marriage as soon as they please. It is not an uncommon thing for a man thus separated from his wife to maintain her in a house near his own, furnish her with the necessary means of subsistence for herself and family, and continue apparently in perfect friendship with her, while at the same time he takes another bride in her place." *

The custom of entering into civil contracts of this kind is of old standing; for Lobo remarked, "A husband that doth not like his wife may easily find means to make the marriage void; and, what is worse, may dismiss the second wife with less difficulty than he took her, and return to the first; so that marriages in this country are only for a term of years, and last no longer than both parties are pleased with each other; which is one instance how far distant these people are from the purity of the primitive believers, which they pretend to have preserved with so great strictness. The marriages are, in short, no more than bargains, made with this proviso that when any discontent shall arise on either side, they may separate, and marry whom they please, each taking back what they brought with them." †

According to Mr. Vivian, provision is sometimes made at the time of the secular marriage for the event of divorce—" a certain sum is agreed upon for payment by the husband. My friend, the coffee-planter at Harrar, had to promise to pay £10 to his little Galla wife if ever he sent her away, but that was considered an exceptionally heavy fee." ‡

There are, however, nuptials of greater sanctity.

* "Life in Abyssinia," p. 268.

† Johnson's translation. ‡ "Abyssinia," p. 232.

Rassam thus described them: "The most binding marriage with the Abyssinians consists of an interchange of vows between the bridegroom and bride, confirmed by their jointly partaking of the holy Eucharist; in fact, the union in this case is solemnized much in the same way as in other Christian Churches. Here, however, as elsewhere, certain breaches of their mutual vows by either party dissolves the tie and renders the transgressor legally obnoxious to punishment; but as Abyssinian law in such matters has been disregarded for centuries, and Might has taken the place of Right, it follows that an offending husband generally escapes with impunity; so also does the guilty wife, if she happens to belong to the family of a powerful chief. Hence, an unprincipled husband, when tired of his wife, finds no difficulty whatever in getting rid of her; and such repudiation is undoubtedly very common. Many cases of the kind fell under my own cognizance, and, in nearly all, the husbands were leading incontinent lives; on the other hand, I never heard of a single instance of a wife who had been sacramentally married proving unfaithful to her husband, even after his repudiation of her. Most women so situated remain single, and many become nuns. In consequence of this deplorable state of things, Abyssinian females generally entertain a great distrust of the opposite sex, and not one in twenty would willingly contract the more binding marriage." *

Mr. Vivian confirmed this observation. "Marriage is not popular with the women," he wrote, "in either of these countries" (Abyssinia and Somaliland), "and they will only consent to it when physical force is actually used." He added, with regard to Abyssinia,

* "The British Mission to Abyssinia," vol. ii. pp. 215, 216.

"The permanent marriage seems to be very rare, only priests and persons of extraordinary piety indulging in it." * It is to be hoped that these worthies do not resort to violence in order to gratify their wish. Plowden said of this form of wedlock, that "it is usual in old age only to take the sacrament together in the church, thereby pledging themselves to fidelity." †

Doubtless this account of the Abyssinian usage is correct, for it agrees with the observations made by Parkyns. "Church marriages," he wrote, "are rarely solemnized except between persons, who, having first been civilly married, and having afterwards lived happily together till the decline of life, begin to feel that they could not hope to suit themselves better, and so determine to sanctify the marriage by going to church and partaking of the Sacrament. The bond is then considered indissoluble." ‡ Stern stated that the Negus is in the same position as the clergy with regard to monogamy.§

"According to the canons of the Abyssinian Church, the King is bound by the same marital laws as a priest; and, consequently, if his wife dies, he dare not marry another. The bereaved predecessors of Theodoros scrupulously evaded such a contingency by substituting the regularly stored harem in the place of the one lawful wife." ||

Rassam made the following observations on the third form of wedlock: "This last is little better than concubinage. The contracting parties merely engage to cohabit during pleasure, and while so living are regarded as husband and wife. The national Church recognizes

* "Abyssinia," p. 231. † Hotten, p. 137.
‡ "Life in Abyssinia," p. 268. § See p. 223.
|| "Wanderings among the Falashas," p. 118.

only the sacramental marriage as valid; but the laity, as a rule, set all ecclesiastical law in such matters at defiance. Hence, a wealthy Abyssinian Christian, who is debarred by the two higher degrees of wedlock from having more than one wife, may nevertheless have as many third-rate wives as he pleases, and cohabit with them simultaneously. In the course of my inquiries into these matrimonial customs, and the laws affecting inheritances among this peculiar people, I applied to the Abouna to aid me in the research. His reply was, 'My son, you have asked me questions which I am unable to answer. This only I can tell you; Abyssinian marriages, with few exceptions, are so abominably revolting that the issue are all bastards.'" *

From the opinions quoted it may be judged that the inclination towards conjugal constancy is usually greater in the wife than in the husband, but this is not invariably the case. Rassam tells the following story of a couple at Magdala. "The husband was in such dread of losing his partner, knowing that, as he had been united to her by the secondary (civil) marriage only, she might leave him any day, that her refusal to accompany him to the altar, there to partake of the Lord's Supper with him, in token of their more indissoluble union, nearly drove him mad. The matter was eventually referred to the Abouna, my intervention being also sought, and after considerable trouble we overcame the obstinacy of the lady, and induced her to be sacramentally joined to her lovesick lord." † Of course this may have been only an instance of an Ethiopian woman's mistrust of the male sex.

Lobo gave some curious details of the penalties that

* "The British Mission to Abyssinia," vol. ii. pp. 219, 220.

† Ib. vol. ii. p. 216.

followed upon infidelity when he was in the country. "They have here a particular way of punishing adultery: A woman convicted of that crime is condemned to forfeit all her fortune, is turned out of her husband's house in a mean dress, and is forbid ever to enter it again. She has only a needle given her to get her living with. Sometimes her head is shaved, except one lock of hair which is left her, and even that depends on the will of her husband, who has it likewise in his choice whether he will receive her again or not. If he resolves never to admit her, they are both at liberty to marry whom they will. There is another custom amongst them yet more extraordinary, which is, that the wife is punished whenever the husband proves false to the marriage-contract; this punishment indeed extends no farther than a pecuniary mulct; and what seems more equitable, the husband is obliged to pay a sum to his wife. When the husband prosecutes his wife's gallant, if he can produce any proofs of a criminal conversation, he recovers, for damages, forty cows, forty horses, and forty suits of clothes, and the same number of other things. If the gallant be unable to pay him, he is committed to prison, and continues there during the husband's pleasure; who, if he sets him at liberty before the whole fine is paid, obliges him to take an oath that he is going to procure the rest, that he may be able to make full satisfaction. Then the criminal orders meat and drink to be brought out, they eat and drink together; he asks a formal pardon, which is not granted at first; however, the husband forgives first one part of the debt, and then another, till at length the whole is remitted." *

According to Mr. Wylde, Abyssinian girls have many attractions for men. "They perfectly understand the

* Johnson's translation.

utility of 'feeding the beast' with a nice dinner to keep him good-tempered, and from what I have seen of these young ladies, they do everything they possibly can to make a man happy, and being good-tempered, jolly girls, they seldom 'nag,' and no wonder the southern Italians have taken a liking for them, and find themselves perfectly happy in their society." He admits that "their great drawback is their dirtiness;" but adds, "all those that get the chance of being clean keep themselves very neat and tidy." However, "nearly all the lower-class Abyssinian women use oil or fat for their heads; this they do to keep the small parasites quiet, as they cannot get about when the head and hair are thickly besmeared and saturated, and the oil or fat also serves for softening the skin of the face and preventing it from chapping in the cold weather, or blistering during the hot season of the year." * Mr. Vivian remarked of the Ethiopian woman, "though her features are comely, she is not the sort of person one would care to choose as a companion. For one thing, I do not suppose that she ever washes herself in her life, the butter on her hair grows rancid and emits a peculiarly pungent odour, which affronts the nostrils when you pass her in the desert, and wherever she goes she carries with her a large black cluster of flies congregated on her back." † Probably she would "find herself" under the influence of a cleaner civilization.

Conceptions of medicine among the Abyssinians have changed for the better during recent years. Harris has given an account of some of the singular notions and practices which prevailed in his day. He took a supply of medical stores into the country,

* "Modern Abyssinia," p. 245 *seq.*

† "Abyssinia," p. 231.

and he and the doctor with the expedition (Assistant-Surgeon Kirk) acted as physicians extraordinary to the King, with the following results :—

"The most particular inquiries were instituted relative to the mode of counteracting the influence of the evil eye, and much disappointment expressed at the unavoidable intimation that the dispensary of the foreigners contained neither 'the horn of a serpent,' which is believed to afford an invaluable antidote against witchcraft: no preservative against wounds in the battle-field, nor any nostrum for 'those who go mad from looking at a mad dog.' 'We princes also fear the small-pox,' said his Majesty, 'and therefore never tarry long in the same place. Nagási, my illustrious ancestor, suffered martyrdom from this scourge. Have you no medicine to drive it from myself?'

"Vaccine lymph there was in abundance, but neither Christian, Moslem, nor Pagan had yet consented to make trial of its virtues. Glasses hermetically sealed, betwixt which the perishable fluid had been deposited, were exhibited, and its use expounded. 'No, no!' quoth the King, as he delivered the acquisition to his master of the horse, with a strict injunction to have it carefully stitched in leather, 'this is *talakh medanit*, very potent medicine indeed; and henceforth I must wear it as a talisman against the evil that beset my forefathers.'

"'You must now give me the medicine which draws the vicious waters from the leg,' resumed his Majesty, 'and which is better than earth from Mount Lebanon; the medicine which disarms venomous snakes, and that which turns the grey hairs black; the medicine to destroy the worm in the ear of the Queen which is ever burrowing deeper; and, above all, the medicine of the seven colours, which so sharpens the intellect as to enable him who swallows enough of it to acquire every sort of knowledge without the slightest trouble. Furthermore, you will be careful to give my people *none of this.*' " *

Plowden, writing of the *defterers,* said: "They are often more learned than the priests, and equally take advantage of the general ignorance. Their principal gain is by writing amulets and charms against every disease, almost against death. . . . They also profess medicine, and as they do not much analyze the effects of their drugs, many an unfortunate falls a victim to some poisonous plant administered as a love philtre. Most of them are hangers-on of the different churches; they are generally cunning, debauched, and mischief-makers." †

Mr. Wylde, speaking of the Abyssinian clergy of the present time, says, "They know that the days are gone by when every one came to them for some charm or a little holy water to cure a complaint; the very practical nineteenth-century doctor is to be found, and not only the congregation has deserted to the modern school of medicine, but the priest himself will trust in the new treatment in preference to running the risk in getting cured by faith or unfiltered holy water." ‡

A peculiar disease is found in Abyssinia, chiefly but not exclusively among the women. Dufton describes a typical case: "I had the satisfaction of seeing in Gaffat a case of *bouda.* This term is given to a phenomenon of mental abstraction, which the natives explain as 'being possessed by the devil.' The case I am about to mention happened to a female in the service of one of

* "The Highlands of Ethiopia," vol. ii. p. 156 *seq.*
† Hotten, p. 166.
‡ "Modern Abyssinia," 1901, p. 143.

the Europeans. Her symptoms began in a kind of fainting fit, in which the fingers were clenched in the palms of the hands, the eyes glazed, the nostrils distended, and the whole body stiff and inflexible. Afterwards she commenced a hideous laugh in imitation of the hyena, and began running about on all fours; she was then seized by the bystanders, and a *bouda* doctor having been called in, this individual began questioning her as to the person who had possessed her with this hyena devil. She said he was a man living in Gooderoo, south of Abyssinia, and also told how long the spirit would be in possession, and what was required to expel him. Great care must be taken of persons thus afflicted, as cases of this kind sometimes end in death. All their demands for dress, food, trifles of any sort, must be strictly attended to. In the height of the frenzy they will sometimes carry out the idea of their hyena identity to such an extent as to attack any animal that may happen to be in the way. One woman fancied she would like a little donkey-flesh; so to gratify her strange taste she seized hold by her teeth to the hinder part of one which happened to be near. Off went the astonished beast at a pace that nothing in the form of persuasion will lead him to adopt for the gratification of man. Off, too, clinging tight with her teeth to his haunches, went the frenzied girl. Only force would induce her to forego the tender morsel.

"They have several cures for this strange attack; but the never-failing one is a mixture of some obscene filth, which is concealed in some part of the house, whereupon the woman is said to go directly, on all fours, to where it is and swallow it. This would seem incredible, but thousands of corroborative facts, known to Abyssinian residents, put it beyond a doubt.

"The power of possessing persons with the devil is attributed mostly to Jewish blacksmiths; and women and children are terrified when they meet, in a solitary place, a blacksmith who is a Jew. These sorcerers are also said to be endued with the power of changing the shape of the object of their incantations." *

Stern has given some interesting particulars with regard to these seizures. "During the rainy season, when the weather, like the mind, is cheerless and dull, the *boudas*, as if in mockery of the universal gloom, celebrate their saturnalia. In our small settlement, the monotony of our existence was constantly diversified by a *bouda* scene. Towards the close of August, when every shrub and tree began to sprout and blossom, the disease degenerated into a regular epidemic; and in the course of an evening two, three, and not unfrequently every hut occupied by natives would ring with that familiar household cry. A heavy thunderstorm, by some mysterious process, seemed invariably to predispose the people to the *bouda's* torturing influence.

"I remember one day, about the end of August, we had a most terrific tempest. . . . The noise and tumult of the striving elements had scarcely subsided, when a servant of Mr. Mayer, a stout, robust, and masculine woman, began to exhibit the *bouda* symptoms. She had been complaining the whole noon of languor, faintness, and utter incapacity for all physical exertion. About sunset her lethargy increased, and she gradually sank into a state of apparent unconsciousness. Her fellow-servants, who were familiar with the cause of the complaint, at once pronounced her to be possessed. To outwit the conjuror, I thought it advisable to try the effect of strong liquid ammonia on the nerves of the

* "A Journey through Abyssinia," p. 167 *seq.*

Evil One. The place being dark, faggots were ignited; and in their bright flickering light we beheld a mass of dark figures squatted on the wet floor around a rigid, motionless, and apparently dead woman. I instantly applied my bottle to her nose; but although the potent smell made all near raise a cry of terror, it produced no more effect on the passive and insensible patient than if it had been water from the newly formed rivulets."

The landlord of the settlement, "an amateur exorcist, almost by instinct, as if anticipating something wrong in that part of his domains occupied by the Franks, made his appearance in the very nick of time. This bloated and limping dotard, who had wasted his youth and manhood in folly and vice, for which, in his old age, he sought to atone by discarding one after the other of his former wives, and by poring over the legends of saints and martyrs, no sooner hobbled into the hut than the possessed woman, as if struck by a magnetic wire, burst into loud fits of laughter and the paroxysms of a raving maniac. Half a dozen stalwart fellows caught hold of her, but frenzy imparted a vigour to her frame which even the united strength of these athletes was barely sufficient to keep under control. She tried to bite, kick, and tear every one within reach; and when she found herself foiled in all those mischievous attempts, she convulsively grasped the unpaved wet floor, and, in imitation of the hyena, gave utterance to the most discordant sounds. Manacled and shackled with leather thongs, she was now partly dragged and partly carried to an open grassy spot; and there under the starry vault of heaven, and in the presence of a considerable number of people, the conjurer, in a businesslike manner, began his exorcising art. The poor sufferer, as if conscious of the dreaded old man's presence, struggled frantically to escape his skill; but the latter, disregarding her entreaties and lamentations, her fits of unnatural gaiety and bursts of thrilling anguish, with one hand laid an amulet on her heaving bosom, whilst, with the other, he made her smell a rag, in which the root of a strong scented plant, a bone of a hyena, and some other abominable unguents were bound up. The mad rage of the possessed woman being instantaneously hushed by this operation, the conjurer addressed himself to the *bouda* (evil spirit), and, in language not fit for ears polite, requested him to give his name." The demon is always masculine, whatever may be the sex of the person possessed. Question and answer followed, and finally the "conjurer" formally exorcised the demon. "'I command thee in the name of the Blessed Trinity, the twelve apostles, and the three hundred and eighteen bishops at the council of Nicaea, to leave this woman, and never more to molest her.'

"The *bouda* did not feel disposed to obey the conjurer; but on being threatened with a repast of glowing coals, which the majority do not relish, he became docile, and in a sulky and ventriloquizing tone of voice, promised to obey the request.

"Still anxious to delay his exit, he demanded something to eat; and to my utter disgust, his taste was as coarse as the torments inflicted on the young woman were ungallant. Filth and dirt of the most revolting description, together with an admixture of water, were the choice delicacies he selected for his supper. This strange fare, which the most niggardly hospitality could not refuse, several persons hastened to prepare, and when all was ready, and the earthen dish had been hidden in the centre of a leafy shrub, the conjurer said to the *bouda*, 'As thy father did, so do thou.' These words had scarcely escaped the lips of the exorcist, when the possessed person leaped up, and, crawling on all fours, sought the dainty repast which she lapped with a sickening avidity and greediness. She now laid hold of a stone, which three strong men could scarcely lift, and raising it aloft in the air, whirled it madly round her head for two seconds, and then fell senseless on the ground. In half an hour she recovered, but was quite unconscious of what had transpired.

"Three other women had similar attacks that same evening, and that, too, without any premonitory symptoms. I tried to deceive one, and instead of the disgusting concoction, put a wooden dish with bread and water in her way; but on smelling it she shrank from its contents, and rapidly crept on till the strong effluvia brought her to the spot where the loathsome viands were concealed. . . .

"Next in importance to the *bouda* is the *zar*. This malady is exclusively confined to unmarried women, and has this peculiar feature, that during the violence of the paroxysm it prompts the patient to imitate the sharp discordant growl of the leopard. I recollect that the first time I saw a case of this description, it gave me a shock which made my blood run cold. The sufferer was a handsome, gay, and lively girl, a little above fifteen. In the morning she was engaged as usual in her work, when a quarrel ensued between her and other domestics. The fierce dispute, though of a trifling character, roused the passions of the fiery Ethiopian to such a pitch that it brought on an hysterical affection. The natives all cried, 'She is possessed,' and certainly her ghastly smile, nervous tremor, wild stare, and unnatural howl, justified the notion. To expel the *zar*, a conjurer, as in the *bouda* complaint, was formerly considered indispensable; but by dint of perseverance the medical faculty of the country, to their infinite satisfaction, have at length made the happy discovery that a sound application of the whip is quite as potent an antidote against this evil as the necromancer's spell." *

Parkyns said that he had seen "above a hundred" cases of *bouda.* He alluded to the superstition with regard to blacksmiths. "In Abyssinia their trade is hereditary, and considered as more or less disgraceful, from the fact that blacksmiths are, with very rare exceptions, believed to be all sorcerers, and are opprobriously called *bouda*. . . .

"Few people will venture to molest or offend a blacksmith, fearing the effects of his resentment. The greater part of the 'possessed' are women; and the reason of their being attacked is often that they have despised the proffered love of some *bouda,* or for other similar cause.

"It is a custom in Abyssinia to conceal the real name by which a person is baptized, and to call him only by a sort of nickname, which his mother gives him on leaving the church. . . . The reason for the concealment of the Christian name is that the *bouda* cannot act upon a person whose real name he does not know. Should he, however, have obtained the true name of his victim, he takes a particular kind of straw, and muttering something over it, bends it into a circle, and places it under a stone. The person thus doomed is taken ill at the very moment of the bending of the straw; and should it by accident snap under the operation, the result of the attack will be that the patient dies.

* "Wanderings among the Falashas," p. 154 *seq.*

"This malady and the Tigritiya are no doubt often purposely counterfeited by servant-maids to evade their work, and by others to excite pity, attract attention, and get themselves pulled about by the men. . . .

"The first case which I ever saw, and which I consequently watched very attentively and noted down, was that of a servant-woman at Rohabaita. The first day she complained of general langour, and of a stupid heavy feeling about the head. Towards evening this seemed to increase, when she cried a little, but was perfectly reasonable, and excused herself by saying that it was only because she felt low and melancholy. An hour after this, however, she burst out into hysterical laughter, and complained of violent pain in the stomach and bowels. It was at this stage that the other servants began to suspect that she was under the influence of the *bouda*. In a short time she became quiet, and by degrees sank into a state of lethargy, approaching to insensibility. Either from excellent acting and great fortitude, or from real want of feeling, the various experiments which we made on her seemed to have no more effect than they would have had on a mesmeric somnambulist. We pinched her repeatedly; but pinch as hard as we could, she never moved a muscle of her face, nor did she otherwise express the least sensation. I held a bottle of strong sal-volatile under her nose, and stopped her mouth, and this having no effect, I steeped some rag in it and placed it in her nostrils; but although I am sure that she had never either seen, smelt, or heard of such a preparation as liquid ammonia, it had no more effect on her than rosewater. She held her thumbs tightly bent inside her hands, as if to prevent their being seen. On my observing this to a bystander, he told me that the thumbs were the *bouda's* particular perquisite, and that he would allow no person to take them. Consequently, several persons tried to open her hand and get at them; but she resisted with what appeared to me wonderful strength for a girl, and bit their fingers till in more than one instance she drew blood. I, among others, made the attempt, and though I got a bite or two for my pains, yet either the devil had great respect for me as an Englishman and a good Christian, or she had for me as her master, for the biting was all a sham, and struck me as more like kissing than anything else, compared with the fearful wounds she had inflicted on the rest of the party.

"I had a string of amulets which I usually wore, having on it many charms for various maladies; but I was perfectly aware that none for the *bouda* was among them. Still, hoping thereby to expose the cheat, I asserted that there was a very celebrated one, and laid the whole string on her face, expecting that she would pretend to feel the effects, and act accordingly; but to my surprise and disappointment, she remained quite motionless. Several persons had been round the village to look for some talisman, but only one was found. On its being applied to her mouth, she for an instant sprang up, bit at it, and tore it, but then laughed, and said it was weak, and would not vex *him*. . . . I deluged her with bucketfuls of water, but could not even elicit from her a start or a pant, an effect usually produced by water suddenly dashed over a person. At night she could not sleep, but became more restless, and spoke several times. She once remarked, in her natural tone of voice, that she was not ill, nor attacked by the *bouda*, but merely wished to return to Adoua. She said this so naturally that I was completely taken off my guard, and told her that of course she might go, but that she must wait till the morrow. The other people smiled, and whispered me that it was only a device of the *bouda's* to get her into the forest, and there devour her.

"Singular coincidences not unfrequently occur in such a way as to encourage superstitious persons in their credulity; and, strange to say, that very night, for the first and last time that I ever heard him during my stay at Rohabaita, the hyena kept howling and laughing close to the village. . . .

"At night I ordered the people to close the door of the hut, and lie across it, some inside and some out. These precautions, however, did not satisfy them; and they insisted on having the young woman bound hand and foot, as the only means of preventing her escape. She lay pretty still, merely moaning, and occasionally starting up when the hyena called. I was lying on a couch, she and the other people on the floor. Determined to see the issue of the affair, I watched her narrowly, and when the guard dropped off to sleep, one by one, I pretended to do so likewise. She also was perfectly still for near an hour, and I fancied that she too had fallen asleep, when suddenly, the hyena calling close by, she, to my astonishment, rose without her bonds, which I had seen, as I imagined, securely fastened. She then crept on all-fours towards the door, which she succeeded in partly opening. I was just going to spring on her when one of the heavily sleeping guards made a noise, which sounded something like a grunt or a snore, and it appeared to me that she stifled a laugh. This led me to believe more and more that she was shamming; but I said nothing, and she returned of her own accord to her place. Although she did not sleep during the whole night, yet she remained still as long as the people were quiet, only moaning a little whenever any one, by yawning or otherwise, showed signs of waking. Next day she appeared a little better, and talked more rationally, but still very wildly, and would neither eat nor drink. Once we allowed her to leave the hut for a moment, on pretence of necessity, and she went quietly enough; but on her return being delayed longer than we considered right, parties were sent out in quest of her, and after a long search she was discovered more than a mile from the hut, and making for the thickest of the jungle.

"The second night was passed much as the former one; but the following day we prevailed on her to take a little bread. On swallowing a piece about the size of a nut she became very sick; and a draught of water produced a similar result. A better night seemed to do her good, for on the following day she managed to eat a little, and by slow degrees recovered her health.

"Since this occurrence I have witnessed many hoaxes easy to discover, but also many which I could never see through, although I tried every method that my small stock of ingenuity could invent.

"I remember once a poor weakly girl on whom I had tried several false charms, but without her moving. She was lying, apparently senseless, in the inner court of a house at Adoua, and numbers of persons were passing to and fro to see her, when, suddenly starting up, she screamed and struggled with so much strength, that, on seizing her by the legs and shoulders, I and three other powerful men could scarcely keep her down. A man near me said, 'I am sure some one has with him a strong charm; if so, let him produce it.' All denied the fact; but just then a stranger entered from the outer court, when she cried out, 'Let me alone, and

I will speak.' This man was an Amhara soldier, perfectly unknown to all of us, but who, hearing one of our people inquire for charms in a house where he was drinking, had volunteered to try one which he wore, and which he declared to be very potent. On placing the amulet near her, she yelled and screamed horribly. The owner (addressing her as a woman) said, 'Will you declare yourself if I take it away?' She howled still more at this insult, as she called it, and said, 'I am no woman.' The question was then repeated several times in the masculine, but she invariably attempted to evade a direct answer, till, as if worn out, she exclaimed, 'I will tell all; only take it far from me!' It was accordingly removed to some distance from her, and she immediately sank down as if exhausted."

When the cure had made some progress, "the *bouda*, anxious to delay his exit as long as possible, demanded food (as he always does) before leaving her. A basin was fetched, in which was put a quantity of any filth that could be found (of fowls, dogs, etc.), and mixed up with a little water and some ashes. I took the basin myself, and hid it where I was positive she could not see me place it, and covered it up with some loose stones which were heaped in the corner. The *bouda* was then told that his supper was prepared, and the woman rose and crawled down the court on all fours, smelling like a dog on either side, till, passing into the yard where the basin was, she went straight up to it, and, grubbing it out from the place where it was hidden, devoured its abominable contents with the utmost greediness. The *bouda* was then supposed to leave her, and she fell to the ground, as if fainting. From this state she recovered her health in a few days.

"I had forgotten to mention that one of the principal

symptoms is a great longing for charcoal, of which the patient will eat any quantity she can obtain. The first proof of the devil's leaving her is her allowing her thumbs to be taken hold of."

On another occasion Parkyns, suspecting a hoax, concocted a charm which was composed of two or three bits of dry bamboo roots wrapped in a piece of paper, an old leaf or two, some pipe-ashes, and a tuft of hair cut from the tail of his dog. Then "proceeding to the place where the sufferer lay, I ordered a large-mouthed jar to be filled with dry mules' dung and lighted. When this had been done, and the smoke began to rise in clouds, I put into it a small quantity of my charm, with every appearance of caution and care, which done, we seized the unfortunate victim, and with some difficulty forced her head close to the jar's mouth, and then rolled a *quarry* (*shama*) round it and the jar, so as at once to keep her fast, and prevent the escape of the smallest quantity of smoke. As may be imagined, in a moment she began to cough violently, and at last, being almost suffocated, cried out, 'Let me off, for pity's sake; I am not ill, but only shamming.' I solemnly asserted this to be only a device of the Evil One to get away from the charm, and held on, till her cries for pity, for the sake of the Virgin, of Oubi, etc., becoming more confused, and her cough more violent, I feared lest she might suffer too much if kept longer.

"That magic vessel was preserved in a conspicuous situation in the hut where the women worked till I left the place; and I must say that the attacks of *bouda* were less frequent after this occurrence than before; though I still had occasional cases where the sunken eye and vacant stare cheated me into tenderness of heart, and I refrained from the use of my singular but most efficacious remedy."

Parkyns has also given an account of a similar disorder called *tigritiya,* which may be the same as the *zar* mentioned by Stern, though from the description, it appears to be more serious. The *tigritiya,* says Parkyns, is the near relation of the *bouda,* "but generally a far more disagreeable and dangerous devil than he. His dependents are supposed to be found in great numbers in Godjam, which province, indeed, is celebrated for sorcerers of all kinds. The first symptoms usually are the gradual wasting away of the attacked person, without any cause being apparent either to herself or her relatives. At last, however, after dieting, and the ordinary medicines have been unsuccessfully tried, it becomes a matter of suspicion to her friends, who determine, on ascertaining, whether or no she be afflicted with this devil. The first thing to be observed is to feed her daintily and dress her neatly. As her complaint and this treatment advance, she becomes peevish and fretful, and is always longing for something or other. Whatever she demands must be procured, else she will become sulky, and, covering up her head, remain sometimes for days without eating or speaking. Ornaments of all kinds require to be borrowed for her, often at much trouble to her unfortunate relations; for she is rarely satisfied unless she gets an assortment of those worn by both sexes, even to the lion's skin of a warrior; and these are frequently almost impossible to procure. With some persons it is necessary to have recourse to music before the real cause of their complaint can be discovered. Drums and other musical instruments are collected outside the chamber door, and the musicians suddenly strike up all together, when the patient is not expecting it. If her illness be of an ordinary kind, she will, of course, beg of them to desist, but, if possessed, she will leap or fall from her couch to the ground, and commence shrugging her shoulders and swinging her head to and fro in time with the music; beginning with a slight movement, but gradually increasing in pace as she appears to become excited, till at last her motions are so violent that a spectator is led to fear for the safety of her neck. It is truly wonderful to see a sick person whom you have just beheld stretched on a bed, a weak, emaciated bag of bones, apparently without strength to rise, keeping up this very fatiguing motion with a velocity and power of endurance that would be astonishing even in an ordinarily strong person. On the music's ceasing she rests, and then begins to speak, telling *his* name (that is, the devil's) and his family; and it is usually after this that she demands the trinkets, specifying each article that she requires, and making their production the condition of her dancing again. She, as I before said, generally contrives to name objects that are most difficult to obtain, such as the silver-ornamented velvet worn only by great war-chiefs; and much sorrow and trouble does this cause to her relations, for she will not be persuaded to show any sign of animation if a single article she has named be not forthcoming; and on her dancing and singing is supposed chiefly to depend her chance of recovery.

"The mode of ejecting the Old Gentleman from his temporary possession is something similar to what we have already described. After the patient has been induced to dance pretty frequently, it is to be hoped that the devil is put into a tolerably good humour, which desideratum can only be obtained by making frequent and polite inquiries when he will be pleased to have music, and by promptly procuring everything he may demand, by the mouth of the possessed, whether it be food or wearing apparel. The auspicious moment arriving, the friends of the sufferer inquire of her when *he* will be disposed to quit his present residence and go to his own place. The first time this question is ventured upon, the spirit generally replies that he is not yet satisfied with the entertainment he has received at their hands, and seizes the opportunity of demanding something more. After a time, however, if things go on well, he fixes a day for taking his leave, usually the seventh after the day on which he makes his announcement. The happy time having arrived, a large party is assembled in some wild spot in the country, and there must be feasting, dancing, and music, as before, but, if possible, carried on with more than usual spirit. The patient joins in with the rest; and the devil, when satisfied, announces his intention to depart. This he signifies through the medium of his victim, who takes off her finery; then, bowing her head, she kisses her hands in token of farewell, and starts off, running at a pace which few men could equal. She cannot, however, keep it up for more than from fifty to a hundred yards, when she falls to the ground senseless. At this moment the spirit is supposed to have left her; and lest he should find himself worse off outside than in, and, changing his mind, return to her, five or six active young men are prepared beforehand, and as soon as she starts they follow her, and, coming up with her just as she is falling, place themselves in threatening attitudes round her body, one holding a drawn sword, another firing a charge of powder out of a matchlock, and a third brandishing a lance. This is supposed to intimidate the fiend, and prevent his return, should he be so disposed.

"The woman, who but a moment before was outdoing all the party, both in the dance and in running, now lies stretched on the ground helpless and emaciated, as if after a long illness. She faintly calls for water, which is given her; and when a little restored, she is asked her name, which she answers correctly; and as a conclusive proof of her freedom from the power of the Evil One, she is requested to repeat the formerly so much dreaded words—'*B'ism Ab, ou Weld, ou Menfus Kouddos*' (in the name of the Father, and of the Son, and of the Holy Ghost). A sheep or a fowl is killed, broiled on the embers, and eaten with bread. The patient's friends partake of this food when the devil goes out of her. The bones and remains of the meat are burned with fire, and the fragments of the bread buried in the ground. These are so left for the devil, that if he should come back to the place he may remain and feed, and not go on and bother the woman. Thus is the cure complete, though it often takes a considerable time to effect it."

The characteristics of the disease have been mentioned fully because they seem to have some bearing on certain Biblical episodes.

Sometimes, in the case of *bouda*, recourse is had to a very simple and direct method of cure. Parkyns wrote:—"Many instances have been related to me, wherein the friends of the possessed, having procured charms of sufficient power to force the spirit to declare his name and residence, but not equal to turning him out, went at once to the place indicated, and, seizing on the blacksmith, brought him forcibly to their home, where, having fed him well, he was commanded to quit his victim, and at the same time his life was threatened, lances being pointed at his breast."* It is not stated whether this procedure is effective.

Mr. Wylde dismissed the demon lightly:—"The *bouda*, or evil spirit, that attacks some of the young women, nearly always ugly virgins or hysterical and plain-looking girls that men will never notice, is to my mind the greatest fraud of all their superstitions. Every one has written about it, and I am afraid that they have drawn a good deal on their imagination, and the missionaries who have visited the country have, perhaps, been quite as bigoted as the people they have tried to describe. I believe that these peculiar fits which the women have, when they do all sorts of filthy things, are nothing more than hysteria; many women even in civilized countries are not responsible for their actions when suffering from these complaints, and people who are inclined to believe in the miraculous take for granted what the ignorant peasantry say. I have seen several young women suffering from the '*bouda*' and a bucket of cold water that I have thrown over them, and a good smacking from my servant, has soon sent the devil away, and the only after effect has been that they have been sulky, because they were not made much of."†

Parkyns was not a missionary, and he was a shrewd and levelheaded observer. His comments are to the point:—"To many of my readers it will doubtless appear, and very naturally, that all these symptoms are impostures resorted to by the pretended sufferers as a means of procuring the borrowed finery, and enjoying the gaieties and festivities which are considered as the means of curing the disorder. This sounds very probable; and I cannot deny that such is my own opinion,

* "Life in Abyssinia," p. 300 *seq.*
† "Modern Abyssinia," p. 385.

though there are still some points which have rather puzzled me. First, as in the *bouda*, the extraordinary talent for acting which they display, and then the fact that the imposture has not been discovered and published centuries ago, but that it is still believed by the very people among whom it has so long been and still daily continues to be practised. How is it possible that a woman, who in her youthful days may have been guilty of such a hoax, should suffer herself to be imposed upon and led into so much trouble and expense by her children afterwards? And yet this is of common occurrence. From this last remark let it not be supposed that young women are the only sufferers; men and women of all ages are liable, though the young of the fair sex are perhaps the most frequently possessed.

"One of my servants, who, by the great anxiety he showed in watching and tending the patients, was evidently convinced of the truth of their sufferings, had himself been attacked many years before, and assured me that after his recovery he had not the slightest recollection of anything that had taken place while the fit was on him, but that his friends had told him all about it. How was it possible that this man, supposing his own illness to have been feigned, could be cheated by the same means?

"Lastly, the most puzzling thing of all is when a person acts sickness to such perfection that Azrael himself is deceived, and, mistaking the feigned malady for a real one, finishes it by seriously taking away the life of the shammer.

"The following case will illustrate what I mean. I had a servant named Bairou, a youth of about nineteen, who, from having been several years in the service of Europeans, had acquired a few of their notions; and, among others, had learnt to ridicule the superstitions of his countrymen. He had a sister who had been ill for several months, and no one knew what her complaint was. At his request I went to visit her more than once, but was unable to do anything for her. The fact is, my doctoring is on a very limited scale; and, as even the most eminent physicians agree that the greatest difficulty is to ascertain what is really the matter with the patient, I stood very little chance with Bairou's sister, who complained of nothing, and showed no marked signs of ailing except in entire prostration of strength, and a rapid falling off of flesh. She gradually got worse, till one day her brother came to me and requested me to lend him my ornaments, and also to beg some more from my friends at the camp. I asked whether he was going to be 'arkee' to some friend about to marry; but he answered with a melancholy smile that he wanted them for his sister; as, having tried everything else, their friends had proposed to see if she were possessed, and he, though not believing in such nonsense, was willing to allow them to try the experiment, lest, if anything happened to her, they should upbraid him afterwards for having caused her death by his obstinacy and incredulity.

"I, of course, quite approved of his determination, and easily succeeded in obtaining the articles required of me. She was dressed up in the borrowed finery as she lay on the couch; and at a signal the musicians outside commenced playing. At the first notes her eyes began to brighten, and, raising herself up for the first time during many days, she swayed her body to and fro for a few moments, after the manner of one possessed; but, becoming quickly exhausted, she sank back, saying, with a faint smile, 'It is too late now!' She repeated these words twice; they were the last I ever heard her utter.

Three hours after she was a corpse. Was this, too, a sham? Or what may it be called? Possibly some freak of her disordered imagination."*

Stern had the tendency to moralize which prevailed at the time when he wrote, but there can be little doubt that he offered a sound opinion in the following passage:—"In bringing this demoniacal subject to a close, I am forcibly reminded of the words, 'Be sure your sin will find you out.' That there is something in these diseases and in their mode of cure which transcends ordinary disorders no one who has stood beside a frantic and agonized patient and wondered at the sudden and more than dramatic transition from raving frenzy to childlike docility can well deny; but without deciding whether it is epilepsy, catalepsy, or hysteria, I am quite sure that fiends and spirits have less to do with the matter than the irregular life and dissolute course which so many pursue."†

It would, indeed, be a matter for astonishment if marked nervous affections did not occur in a country where, among the people of both sexes, unwholesome diet and parasitical irritation so often combine their results with those of sensual habits of body and mind acquired early in life.

It has already been said that the Abyssinian soldier wears no uniform, and is not easily distinguished by his appearance from the civilian. But there is a distinct military class, with characteristics of its own. "There is no harder worker," said Mr. Wylde, "than the Abyssinian peasant, and no more harmless and hospitable person when left alone and properly treated‡; and no more truculent, worthless, conceited, lazy, and

* "Life in Abyssinia," p. 318 *seq.*

† "Wanderings among the Falashas," p. 161.

‡ Compare Consul Plowden's remarks quoted on p. 70 of this book.

useless individual than the Abyssinian soldier, who formerly did nothing but prey upon the defenceless cultivator. Circumstances are now altering this, but before the country settles down to modern civilization and makes any great strides forward, a civil revolution must take place, which may not be far distant." * The same writer has pointed out that the soldiery "were called into existence by Abyssinia being surrounded by Mohammedan enemies, and little by little they increased and multiplied till they got out of all proportion to the wants of a peaceful country." The king's exchequer cannot provide pay for these men, so, to keep them quiet, they are allowed to live by exaction. Mr. Wylde's strictures apply to the "lazy, loafing lot of mercenaries who have never done anything in their lives except fighting and looting, men without homes and without territory, ready to fight for those who give the highest pay, and who do not value the lives of their fellow-Christians at the price of a sheep or a jar of hydromel;" and he adds, "I am writing only of the mercenary soldier whose father and grandfather, perhaps, were the same, and not of the bulk of the fighting force of the country, who are yeomen farmers and their servants, or the peasants and their families." †

The state of feeling between the peasantry and the soldiery might have serious consequences if the Abyssinian monarch undertook an important campaign outside his territory. "The great danger to an unpopular king attempting such an expedition would be, in the absence of the army, a rising of an oppressed peasantry, backed up by some European Power, to put down the military party. The arming of the peasantry and farmer class with modern weapons has not altogether

* "Modern Abyssinia," p. 3. † Ib., p. 165.

been a blessing to the present ruler, and may end not only in his downfall, but in that of the barons as well." *

The fighting value of the Abyssinian army is by no means a *quantité négligeable* in the politics of Eastern Africa, and it would be very easy to underrate it. Mr. Wylde, in his narrative of the battle of Adoua, has given a very complete account of the numbers, armament and temper of the Negus's forces, and he does not agree with the estimate formed by Mansfield Parkyns of their personal valour.† He has also described the tactics which they customarily adopt in attacking, and has made some interesting observations on the employment of artillery, mounted infantry, and cavalry in the country.‡

The limitations to the activity of the forces are chiefly determined by commissariat difficulties. "The Abyssinian hordes are the same as the locust, they live on what they can get from the surrounding country; and when they have devoured everything, they have to move on to another place where supplies are procurable. At the outside an Abyssinian who is not one of the regular soldiers can keep the field for a couple of months, and then he has to take one transport animal with him, with a boy or girl, generally the latter, to look after his riding animal and to cook his food. The regular troops are the same; they have to bring supplies with them, which they get from their leaders before they set out on the campaign; after these are finished, unless fresh supplies come forward, they have to live on the country." §

* "Modern Abyssinia," p. 224.

† Compare, *e.g.*, his remark on p. 223, "They have any amount of pluck," with Parkyns's comments quoted on p. 24 of this book.

‡ *E.g.* p. 222. § Ib., p. 216.

Probably the most recently published paper dealing with the Abyssinian army is the despatch, dated June 2, 1904, of Colonel Rochfort, C.B., R.A., included in the official account of the operations in Somaliland.* Colonel Rochfort accompanied the forces which his Majesty Menelek II., on the invitation of the British Government, sent in 1903, "to intercept the Mullah's retreat, should he attempt to escape to the south or west." As might be imagined from the nature of the document, it contains no descriptive account of the Abyssinian Expedition, and very little critical comment upon it. But the following particulars may serve to convey some idea of Abyssinian operations in the field at the present time. The strength of the force first employed was about five thousand men, nearly all of whom were mounted on mules. It left Harrar for its destination (the Webi Shebeli, near Hilowen) on February 18, 1903, under the command of Fituari Gabri.

Surprise parties were sent out to visit all wells within reach to the north, and "the Abyssinians carried out these raids with considerable dash and some success, showing their mobility by the ease with which they covered long distances, sometimes a hundred miles in forty-eight hours; but as at this time they had no special arrangements for carrying water, their scope was necessarily restricted.

"During these operations the main body was attacked by a considerable number of Dervishes. The attack was delivered in thick bush on the arrival of the force in bivouac, and before the zareba was formed; it was pushed home with resolution, but repulsed with loss. The pursuit was carried out with vigour, and the Dervish casualties were estimated at three hundred. The

* Eyre and Spottiswoode.

Abyssinians returned their loss at thirty-one killed and wounded. The Abyssinians were now living on the country, and exigencies of supply rendered it necessary to move down stream to Mekunna. . . .

"On the morning of May 28 the Abyssinians left Hahi, and after a series of forced marches surprised the Habr Suliman section of the Bagheri tribe on the morning of May 31 at Jeyd, which is thirty miles south of Bur. The Dervishes suffered heavy losses, and all their camels and stock were captured. . . . This terminated the first phase of the operations, and the force returned to Harrar."

Later, the Negus consented to send another expedition, and "expressed the wish for an increase in the number of medical officers to accompany his force, and also suggested the issue of water-bottles to his men. At the conclusion of the first phase of the operations I had represented to the War Office that the absence of any special arrangements for carrying water had seriously affected the mobility of the Abyssinians; this want was now remedied by the issue of one thousand twelve-gallon tanks; pumps and waterproof sheets for improvising drinking-troughs were also supplied by his Majesty's Government.

"There is no organized system of supply in the Abyssinian army, and hitherto arrangements made individually to carry one month's supply on a mule or donkey had been found sufficient to enable expeditions to reach a district from which supplies could be drawn; such a course did not meet the conditions under which the present force was about to act, owing to the distance to be covered, the necessity for carrying water, and the total absence of any grain, either on the road or in the probable zone of operations. After some unavoidable delay, arrangements were improvised for carrying two months' supply,* and his Highness Ras Makunnan,† who personally interested himself in the despatch of the force, subsequently supplemented this supply by sending three small convoys as transport became available. The force commenced concentration in the neighbourhood of Harrar on November 27, 1903. . . . The advanced party reached Wardair on January 12, 1904. . . . The resources of the country in the neighbourhood of the River Fafan and the arrival of two more small convoys enabled the force to keep the field until March 28, 1904, when the Abyssinian commander (Fituari Gabri) was advised that the retreat of the Mullah through the Sorl, pursued by the British columns, rendered the presence of the Abyssinian expedition no longer necessary, and the force marched for Harrar." ‡

Those who wish to acquaint themselves with the home life of the Abyssinian people should read the works of Mansfield Parkyns and Mr. Augustus Wylde. The latter has given especially full details of the surroundings and habits of the middle and lower classes of the nation.

Mr. Wylde has chapters on "Shooting in Abyssinia" and "Outfit and Rifles." His book also contains an appendix on "Animals met with in Abyssinia and on the Borders," ranging in its scope from lions to sand-martins. There are several passages dealing with sport and natural history in Parkyns's volume, and Mr. Vivian's work contains observations on the same subjects.

Stecker's scientific notes have been frequently

* Cp. p. 308.

† Mr. Wylde describes the Ras as "by far the cleverest and most enlightened man that the country possesses." He is a possible successor to the Abyssinian throne.

‡ Colonel Rochfort's Report.

mentioned. His brief but pithy treatise amply repays study. Harris's "Highlands of Ethiopia" has valuable appendices on the climate, geology, botany, and zoology of the southern provinces (Vol. II).

Probably no country at the present time offers a better field for research, sport, and exploration than Abyssinia.

APPENDIX

Abyssinia is a deeply interesting country from the point of view of geographical distribution, and it is much to be regretted that Dr. A. J. Hayes did not have the opportunity of collecting insects on a large scale. The animals of the southern half of Arabia are Ethiopian in character; but in the Abyssinian mountains we may expect to find, and we do find, a certain amount of Oriental affinity.

The valuable little collection of insects made by Dr. Hayes has been presented by him to the Hope Department of the Oxford University Museum, where the specimens can be seen and studied by every naturalist interested in the great problems of distribution. The attention of the donor was directed to the Oxford Museum by Mr. W. L. S. Loat, who has himself contributed a large amount of valuable material. Dr. Hayes' collection was made, in February 1903, in the vicinity of Lake Tsana, at a height of about 6500 feet. A complete list is furnished below. Dr. Dixey has kindly determined and made remarks upon the *Pierinae*.

LEPIDOPTERA.

NYMPHALIDAE.

Danainae: 1 *Limnas chrysippus* (Linn.) ♀. The ground colour of the pale tint characteristic of Oriental specimens and usually replaced by a much darker shade in African.

NYMPHALIDAE (*continued*).

DANAINAE: 2 *L. chrysippus* (Linn.) var. *alcippus* (Cram.) ♂ ♂. Typical.

NYMPHALINAE: 1 *Neptis agatha* (Cram.).

1 *Precis cebrene* (Trim.).

PAPILIONIDAE.

PIERINAE: 1 *Catopsilia florella* (Fabr.) ♂.

2 *Colias electra* (Linn.) ♂ ♀.

3 *Terias brigitta* (Cram.) ♂ ♂ ♀.

Dry season forms; not extreme.

3 *Eronia leda* (Boisd.) ♂ ♀ ♀.

One of these females has an orange apical patch on the forewing, almost as distinct as that of the male.

1 *Pinacopteryx* sp.?

A female, rather worn; simulating *Mylothris agathina* ♀.

Probably a new species, but being in poor condition and a single specimen it would not be advisable to describe it.

1 *Belenois severina* (Cram.) ♀. Dry season form.

1 *Phrissura* sp. ♂.

A male, of the *P. sylvia* group. This form of *Phrissura* has not previously been recorded from any part of East Africa.

PAPILIONINAE: 8 *Papilio demodocus* (Esp.).

HYMENOPTERA.

1 *Dorylus fimbriatus* (Shuck.) ♂.

COLEOPTERA.

LAMELLICORNIA.

SCARABAEIDAE: 1 *Oniticellus inaequalis* (Reiche).

Only known from Abyssinia.

CETONIIDAE: 1 *Pachnoda abyssinica* (Blanch.).

1 *Pachnoda stehelini* (Schaum).

Both Abyssinian species.

PHYTOPHAGA.

CASSIDIDAE: 1 *Aspidomorpha punctata* (Fab.).

HETEROMERA.

CANTHARIDAE: 2 *Mylabris*, probably a new species.

NEUROPTERA.

1 *Nemoptera*, probably a new species.

ORTHOPTERA.

ACRIDIIDAE: 1 *Cyrtacanthacris* sp.
1 *Phymateus brunneri?* (Bolivar).
1 *Phymateus leprosus* (Fab.).
1 *Petasia anchoreta* (Bolivar).

MANTIDAE: 1 *Sphodromantis bioculata* (Burm.).
1 *Chiropus aestuans?* (Sauss.).

In addition to the above, Dr. Hayes presented three insects captured by him at Gedaref in the Soudan, including a pair of a magnificent new species of Buprestid beetle of the genus *Sternocera*, taken *in coitu*. This species has recently been described, from Dr. Hayes' specimen and two others in the British Museum, by Mr. C. O. Waterhouse, who has given it the name *Sternocera druryi* ("Ann. Mag. Nat. Hist." Oct., 1904, p. 247). The third insect is an example of a Cantharid beetle, which does great damage to the crops at Gadarif. Its determination as *Mylabris hybrida* (Bohem.) is therefore a matter of some importance.

THE END

PRINTED BY WILLIAM CLOWES AND SONS, LIMITED, LONDON AND BECCLES.

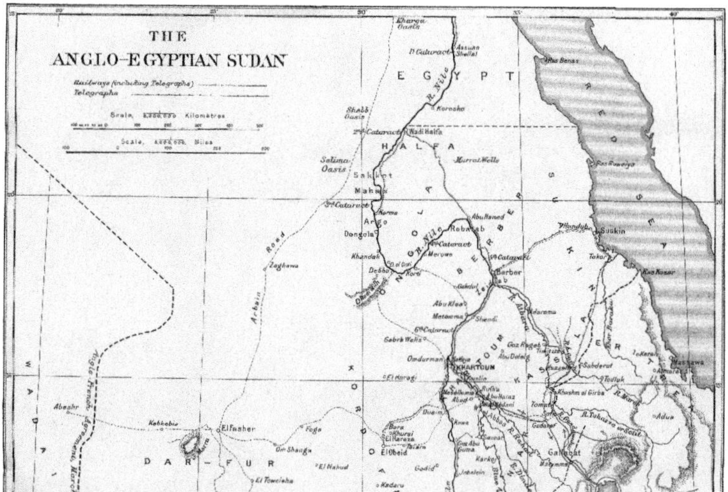

By permission of the Egyptian Government.

London: Smith, Elder & Co.

Stanford's Geog^l. Estab^t. London.

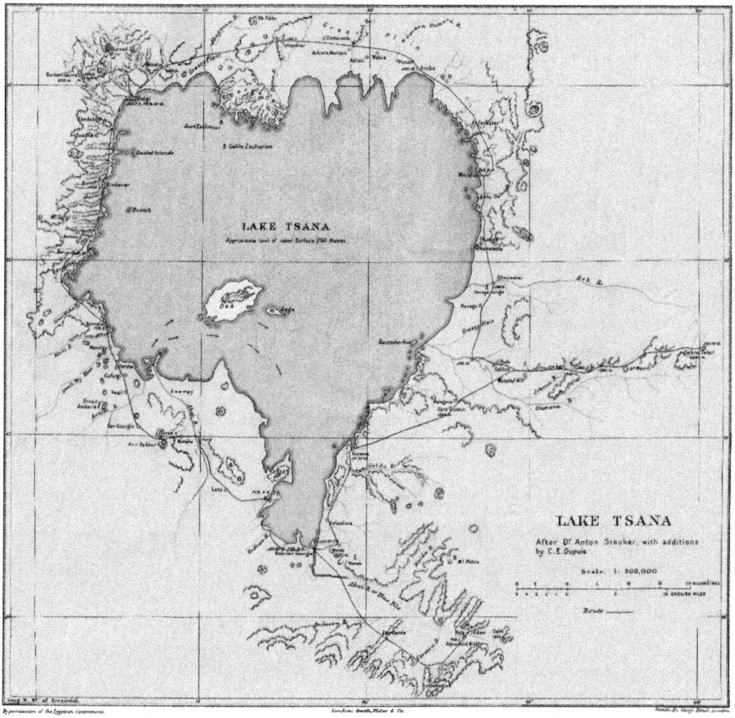

By permission of the Egyptian Government. London: Smith, Elder & Co. Stanford's Geog.l Estab.t London.

Printed in Dunstable, United Kingdom

91223683R00228